# THE HOLODECK
## A SPECIFICATION

# MICHAEL CLORAN

To order additional copies of this book, contact:
Xlibris
0800-056-3182
www.xlibrispublishing.co.uk
Orders@ Xlibrispublishing.co.uk

ISBN:   Softcover       978-1-9845-9275-0
        Hardcover       978-1-9845-9273-6
        EBook           978-1-9845-9274-3

Library of Congress Control Number:      2019919335

Print information available on the last page

Rev. date: 02/07/2020

# Preface

This book has been formed out of my original research on photonic computer hardware design where I scripted on a whiteboard the operation and theory of a theoretical five-bit photonic computer. After some more research was done, this design was considered by me to be out of date. The tutorial was only a glimpse of what could be possible and had many mistakes due to human error. So I made a decision to design the software to simulate photonic computer hardware algorithms at a proof of concept level. This software project took about five years of design and programming, and now I can simulate photonic hardware at a proof of concept level with a full-blown graphical user interface at a motherboard level, and this software was used to aid in the writing of this book.

As my research path unfolded, I acquired a large set of books on hardware and software design, but I also wanted to include volumetrics as a simulation capability. I also wanted to include theoretical touch capability; thus the sections of the book on holodeck theory were formed (as well as the appropriate title for the book) where I had to research material structure at a molecular level. Hopefully, my theory on how lighting and touch could work in a holodeck environment will hold true.

Also note that with this book project, the hardware algorithms are only at stage 1 of the design pipeline, which is proof of concept level, and the mathematics used is high-level theoretical, as the goal of the book is to show a photonic computer hardware algorithm and how it would work at a proof of concept level. Then once this is achieved, it would show simple volumetrics and introduce holodeck theory, the theory of operation of a holodeck, and how some theoretical concepts could be achieved.

# Author

Michael Cloran has a technician diploma in electrical engineering from Kevin Street DIT and a BEng in telecommunications from Dublin City University (DCU). Ever since leaving DCU in 2005, Michael has researched, as a hobby, optical computer design, programming, and 3D graphics. This led to holodeck design theories and concepts and this book.

# Acknowledgements

I would like to take this opportunity to thank my publisher Xlibris for their patience, as this book took a long time to get together. I would also like to thank all those who supported me over those long years.

# Contents

I.      Introduction
II.     List of Figures/Images
III.    List of Tables

**Part 1 The Theoretical Functions of a Photonic Computer**
1.1. Introduction                                                    2
1.2. Optical Switches                                                5
**Processor Functions**
1.3. Shift Left Register                                             6
1.4. Shift Right Register                                            9
1.5. Arithmetic Shift Right                                          11
1.6. Circular Shift Left                                             12
1.7. Circular Shift Right                                            15
1.8. Full Adder                                                      17
1.9. Adder-Subtractor                                                20
1.10. Signed Multiplier                                              23
1.11. Signed Division                                                25
1.12. Floating Point Add/Subtract                                    29
1.13. Floating Point Signed Multiply                                 35
1.14. Floating Point Signed Divide                                   38
1.15. Logic Unit                                                     41
1.16. Two-Bit Magnitude Comparator                                   44
**Optical Core Theory**
1.17. ALU Core Theory                                                52
Introduction to Parallelism within a Core                           53
1.17.1. Threading                                                    54
**Holodeck Motherboard Theory**

1.18.1. Data Centre Holodeck Motherboard Rough Requirements         60
1.18.2. Personal Holodeck Ppv Motherboard Rough Requirements        64
**Part 2.1. Instruction Set Theory**

2.1.0. Instruction Set Theory                                        68
2.1.1. Instruction Set Listing for RISC-V                            68
2.1.2. Theory and Notes                                              76
**Part 3. Volumetric Theory of Operation**

3.1. Introduction                                                    88
3.2. Basic Volumetrics                                               88
3.3. Advanced Volumetrics                                            90
3.4. Holodeck Volumetrics                                            91
3.5. Holodeck Lighting Theory                                        95
3.6. Holodeck Solidness Theory                                       96

# Part 4.1. Holodeck Theory of Operation

4.1.0. Architecture Overview   99
4.1.1. Architecture Mappings from Volumetric Space to Motherboard Thread Space   99
4.1.2. Scene   104
4.1.2.1. Basic Scene   105
4.1.2.2. Light   106
4.1.2.3. Terrain   110
4.1.2.4. Sound   113
4.1.2.5. Touch   114
4.1.2.6. Different Materials   116
4.1.2.7. Different Textures   117
4.1.2.8. Different Colours   118
4.1.2.9. Resolution of Force Points   119
4.1.2.9.1. How to Change Update Frequency   121
4.1.2.10. How a Scene Reacts to Movements of Characters/People   122
4.1.2.11. Scrolling Scene   122
4.1.2.11.1. Light   123
4.1.2.11.2. Terrain   124
4.1.2.11.3. Sound   125
4.1.2.11.4. Touch   126
4.1.2.12. Bounds   127

# Networked and Multi-User Holodeck

4.2.1. Introduction   130
4.2.2. Architecture Overview   131
4.2.3. Scene on Main Node   135
4.2.4. Multiple Ppv Nodes   135
4.2.5. Ppv Physical Space to Virtual Space   136
4.2.6. Personal Physical Volumes in More Detail   137
4.2.7. Network Capability   138
4.2.8. Multiple Characters   138
4.2.9. Interaction with Environment   139
4.2.10. Level of Detail   139
4.2.11. Sound at a Point   139
4.2.12. Sound Shield   140
4.2.13. Light Shield   140

## Holodeck Concepts

4.3.1. Basic Laws of Physics   142
4.3.2. Basic Airflow/Wind   143
4.3.3. Basic Rain   144
4.3.4. Basic Snow   144
4.3.5. Character Modelling   144
4.3.5.1. Clothes for Static and Dynamic Scenes   144
4.3.5.2. Character Appearance for Static and Dynamic Scenes   145
4.3.6. Rigging a Character in Volumetric Space   145
4.3.7. Animation   145

4.3.8. Computer Characters    145

4.3.9. Real Characters    146

4.3.10. Weight    146

4.3.11. Safety Protocols and No Safety Enabled    146

**Conclusion**    149

**Appendices**

Appendix A    151

Appendix B    187

**References**    290

# List of Figures

## 1.1. Introduction

Fig.1.1.1. The concept of an electromagnetic wave, part 1          2
Fig.1.1.2.An electromagnetic wave          3
Fig.1.1.3. An intensity modulated wave          4

## 1.3. Shift Left Register

Fig.1.3.1.Shift left register operation          6
Fig.1.3.2.Shift left register initial set-up          6
Fig.1.3.3.Shift left register one bit slot shift left          7
Fig.1.3.4.Shift left register two bit slot shifts left          7
Fig.1.3.5.Shift left register three bit slot shifts left          8
Fig.1.3.6.Shift left register four bit slot shifts left          8

## 1.4. Shift Right Register

Fig.1.4.1 shift right register operation          9
Fig.1.4.2.Optical multimode shift right register being loaded          9
Fig.1.4.3.Optical multimode shift right register shifted by one bit slot          10
Fig.1.4.4.Optical multimode shift right register shifted by two bit slots          10
Fig.1.4.5.Optical multimode shift register shifted by three bit slots          11

## 1.5. Arithmetic Shift Right

Fig.1.5.1.Arithmetic shift right operation          11

## 1.6. Circular Shift Left

Fig.1.6.1.Circular shift left operation          12
Fig. 1.6.2. Circular shift left register initial set-up          12
Fig. 1.6.3. Circular shift left register after one bit shift left          13
Fig. 1.6.4. Circular shift left register after two bit shifts left          13
Fig. 1.6.5. Circular shift left register after three bit shifts left          14
Fig. 1.6.6. Circular shift left register after four bit shifts left          14

## 1.7. Circular Shift Right

Fig. 1.7.1. Circular shift right operation          15
Fig. 1.7.2. Circular shift right register initial set-up          15
Fig. 1.7.3. Circular shift right register with one bit shift right          16
Fig. 1.7.4. Circular shift right register with two bit shifts right          16
Fig. 1.7.5. Circular shift right register with three bit shifts right          16
Fig. 1.7.6. Circular shift right register with four bit shifts right          17

## 1.8. Full Adder

Fig. 1.8.1. Full adder doing 1 + 0 + 0 = 1, decimal 1          18
Fig. 1.8.2. Full adder doing 1 + 1 + 0 = 10, decimal 2          18
Fig. 1.8.3. Full adder doing 1 + 1 + 1 = 11, decimal 3          18
Fig. 1.8.4. A 4-bit optical full adder showing 2 + 3 = 5.          19

## 1.9. Adder subtractor

Fig. 1.9.1. Adder subtractor with mode set to addition adding 1 + 1 = 2 (sum = 0, carry = 1)                                                                                                  20
Fig. 1.9.2. Adder subtractor with mode set to subtract doing the sum 1 − 1 = 0          21
Fig. 1.9.3. Adder subtractor in mode addition showing sum 1 + 1 = 2 (10 in binary)     21
Fig. 1.9.4. Adder subtractor with mode set to subtract with sum 1 − 1 = 0, where the carry can be discarded                                                                                  22

## 1.10. Signed Multiplier

Fig. 1.10.1. Hardware showing M, A, Q, and Q1 registers initial set-up with wavelengths and intensity levels. The counter for 8-bit registers is set initially to 8.                        23
Fig. 1.10.2. Booth's algorithm simulation results                                      24

## 1.11. Signed Division

Fig. 1.11.1. Signed division concept block diagram                                     25
Fig. 1.11.2. Control flow chart of block diagram Fig. 1.11.1                            26
Fig. 1.11.3. Simulation results for signed division for −7 / 2 = −3, remainder −1       28

## 1.12. Floating Point Add/Subtract

Fig. 1.12.1. Flow chart                                                                29
Fig. 1.12.2. Flow chart                                                                30
Fig. 1.12.3. Flow chart                                                                31
Fig. 1.12.4. Flow chart                                                                32
Fig. 1.12.5. Flow chart                                                                33

## 1.13. Floating Point Signed Multiply

Fig. 1.13.1. Flow chart                                                                35
Fig. 1.13.2. Flow chart                                                                36
Fig. 1.13.3. Flow chart                                                                37

## 1.14. Floating Point Signed Divide

Fig. 1.14.1. Flow chart                                                                38
Fig. 1.14.2. Flow chart where the normalised result is the same as before.             39

## 1.15. Logic Unit

Fig. 1.15.1. AND gate array                                                            41
Fig. 1.15.2. OR gate array                                                             42
Fig. 1.15.3 EXOR gate array                                                            43
Fig. 1.15.4 NOT gate array                                                             44

## 1.16. Two-Bit Magnitude Comparator

Fig. 1.16.1. Magnitude comparator, A < B circuit                                       49
Fig. 1.16.2. Magnitude comparator for A = B                                            49
Fig. 1.16.3. Magnitude comparator for A > B                                            50
Fig. 1.16.4. A four-block model magnitude comparator                                   51

## 1.17. ALU Core Theory

Fig. 1.17.0. ALU basics                                                                52

## Introduction to Parallelism within a Core

Fig. 1.17.1. Basic pipelining                                                          53

## 1.17.1. Threading
Fig. 1.17.1.1. The concept of points within a voxel      55
Fig. 1.17.1.2. Basic core layout showing threads and the functions within the threads (branch unit, floating point unit, integer unit and load/store unit)      56
Fig. 1.17.1.3. Core with RAID-controlled optical discs.      59

## 1.18.1. Data Centre Holodeck Motherboard Rough Requirements
Fig. 1.18.1.1. Virtual data centre scene showing a subcell grid unit in voxels      61
Fig. 1.18.1.2. Mappings of cores in subcell grid unit      62
Fig. 1.18.1.3. Data centre concept of a layered motherboard of subcell grid units      62
Fig. 1.18.1.4. Multi-user holodeck concept diagram      63

## 1.18.2. Personal Holodeck Ppv Motherboard Rough Requirements
Fig. 1.18.2.1. Ppv display showing concept of subcell grid unit      64
Fig. 1.18.2.2 Ppv concept of layered motherboard      65

## 2.1.2. Theory and Notes
Fig. 2.1.2.3. Shift left      77
Fig. 2.1.2.4. Shift right      78
Fig. 2.1.2.5. Arithmetic shift right      78
Fig. 2.1.2.6. Add      79
Fig. 2.1.2.7. Subtract      79
Fig. 2.1.2.8. Signed multiply      80
Fig. 2.1.2.9. Signed division      80
Fig. 2.1.2.10. Floating point add/subtract      81
Fig. 2.1.2.11. Floating point signed multiply      81
Fig. 2.1.2.12. Floating point signed divide      82
Fig. 2.1.2.13. Logic unit      82
Fig. 2.1.2.14. Magnitude comparator      83

## 3.2. Basic Volumetrics
Fig. 3.2.1. The concept of a volumetric display      88
Fig. 3.2.2. The concept of a spatially modulated light wave      89
Fig. 3.2.3. The concept of a light wave spatially modulated to a point, and that point being a voxel which has a view cone in the direction of the light wave      89

## 3.3. Advanced Volumetrics
Fig. 3.3.1. The concept of projecting a voxel from several different angles to widen the view capability of the voxel      90

## 3.4. Holodeck Volumetrics
Fig. 3.4.1. Voxel with force points      91
Fig. 3.4.2. Voxel showing three-point moment triangulation. Possible reflection model.      92
Fig. 3.4.3. Voxel showing concept of thin line and thicker line      93
Fig. 3.4.4. Simulation of artificial material showing light and shadow      94
Fig. 3.4.5. Ray of light being reflected at colour but showing unwanted components      95

## 4.1 Holodeck Theory of Operation
## 4.1.1. Architecture Mappings from Volumetric Space to Motherboard Thread Space
Fig. 4.1.1.1. Cell or personal physical volume      99
Fig. 4.1.1.1A. A voxel showing voxel length      100

Fig. 4.1.1.2. Volumetric display 101

Fig. 4.1.1.3. Volumetric display showing cell and voxel/particle block 102

Fig. 4.1.1.4. Motherboard block layout showing mappings to threads 103

**4.1.2 Scene**

Fig. 4.1.2.1a. Cell/Ppv where person is kept near center of the physical space 104

Fig. 4.1.2.1b. Virtual space where a person could be anywhere within scene 105

**4.1.2.1. Basic Scene**

Fig. 4.1.2.1.0. Scene with ambient light and a point light 105

**4.1.2.2. Light**

Fig. 4.1.2.2.1. Scene with ambient light 106

Fig. 4.1.2.2.2. Spatial light modulator showing voxel and view cone 107

Fig. 4.1.2.2.3. Simulated point light with several spatially modulated voxels and their view cones 108

Fig. 4.1.2.2.4. Spotlight in 2D showing view cones 109

Fig. 4.1.2.2.5. Spotlight in 3D showing view cones 109

**4.1.2.3 Terrain**

Fig. 4.1.2.3.2 Reflection 110

Fig. 4.1.2.3.3. A 2D and 3D view of a basic scene with two observers 111

Fig. 4.1.2.3.4. The eye 112

**4.1.2.4. Sound**

Fig. 4.1.2.4.1. Virtual speaker concept 113

**4.1.2.5. Touch**

Fig. 4.1.2.5.1. Force points 114

Fig. 4.1.2.5.2. Voxel showing reflection from simulated surface 115

**4.1.2.6. Different Materials**

Fig. 4.1.2.6.1. Cross section of tree 116

Fig. 4.1.2.6.2. Holodeck cross section of simulated tree 116

**4.1.2.7. Different Textures**

Fig. 4.1.2.7.1. Holodeck simulation of hatchet hitting simulated tree 117

**4.1.2.8. Different Colours**

Fig. 4.1.2.8.1. Optical spectrum 118

**4.1.2.9. Resolution of Force Points**

Fig. 4.1.2.9.1. Simulation of less dense material concept diagram 119

Fig. 4.1.2.9.2. Simulation of more dense material concept diagram 120

**4.1.2.9.1. How to Change Update Frequency**

Fig. 4.1.2.9.1.1. Ray of light reflecting within a voxel 121

**4.1.2.11. Scrolling Scene**

Fig. 4.1.2.11.1. Concept of scrolling scene 122

**4.1.2.11.1. Light**

Fig. 4.1.2.11.1.1. Concept of tile with spatial light modulator spatially modulating to the outer surface of the tile/cell in order to create the scene with these light points per voxel on the surface creating the light for the scene 123

**4.1.2.11.2. Terrain**

Fig. 4.1.2.11.2.1. Concept of centering the user within the tile 124

Fig. 4.1.2.11.2.2. Concept of climbing within a tile/cell 124

Fig. 4.1.2.11.2.3. Concept of walking up a cliff within a tile/cell 125

**4.1.2.11.3. Sound**
4.1.2.11.3.1. Concept of virtual sound from within scene                           125
**4.1.2.11.4. Touch**
Fig. 4.1.2.11.4.1. A wall in a scene with a newton resistance force               126
Fig. 4.1.2.11.4.2. Water in a scene with a newton resistance to movement force   126
**4.1.2.12. Bounds**
Fig. 4.1.2.12.1. Hatchet being swung to penetrate a block of wood                 127
Fig. 4.1.2.12.2. Hatchet cut the block of wood into two pieces                    128
**4.2.2. Architecture Overview**
Fig. 4.2.2.1. Person physical volume in detail                                    131
Fig. 4.2.2.2. Volumetric display complex for 1000 users                           133
Fig. 4.2.2.3. Networked scene with five networked users as example                134
**4.2.3. Scene on Main Node**
Fig. 4.2.3.1. Basic scene in virtual space on the main simulation node            135
**4.2.4.Multiple Ppv Nodes**
Fig. 4.2.4.1. Volumetric display with two people in it                            136
**4.2.5. Ppv Physical Space to Virtual Space**
Fig. 4.2.5.1. Volumetric display with two Ppvs and a walkway in the volumetric
physical space                                                                    137
**4.2.8. Multiple Characters**
Fig. 4.2.8.1. Multiple people and multiple computer characters within the scene   138
**4.2.11. Sound at a Point**
Fig. 4.2.11.1. Sound at a point in 3D space concept diagram                       139
**4.2.12. Sound Shield**
Fig. 4.2.12.1. Concept diagram of soundproofing a Ppv                             140

**Appendix A**
Fig. A.1. Two-input AND gate                                                      151
Fig. A.2. Three-input AND gate                                                    151
Fig. A.3. Four-input AND gate                                                     152
Fig. A.4. Five-input AND gate                                                     152
Fig. A.5. Six-input AND gate                                                      153
Fig. A.6. Seven-input AND gate                                                    153
Fig. A.7. Eight-input AND gate                                                    154
Fig. A.8. Two-input NAND gate                                                     154
Fig. A.9. Three-input NAND gate                                                   155
Fig. A.10. Four-input NAND gate                                                   155
Fig. A.11. Five-input NAND gate                                                   156
Fig. A.12. Six-input NAND gate                                                    156
Fig. A.13. Seven-input NAND gate                                                  157
Fig. A.14. Eight-input NAND gate                                                  157
Fig. A.15. Two-input NOR gate                                                     158
Fig. A.16. Three-input NOR gate                                                   158
Fig. A.17. Four-input NOR gate                                                    159
Fig. A.18. Five-input NOR gate                                                    159
Fig. A.19. Six-input NOR gate                                                     160
Fig. A.20. Seven-input NOR gate                                                   160
Fig. A.21. Eight-input NOR gate                                                   161

Fig. A.22. Two-input OR gate                                  161
Fig. A.23. Three-input OR gate                                162
Fig. A.24. Four-input OR gate                                 162
Fig. A.25. Five-input OR gate                                 163
Fig. A.26. Six-input OR gate                                  163
Fig. A.27. Seven-input OR gate                                164
Fig. A.28. Eight-input OR gate                                164
Fig. A.29. Two-input EXOR gate                                165
Fig. A.30. Three-input EXOR gate                              165
Fig. A.31. Four-input EXOR gate                               166
Fig. A.32. Five-input EXOR gate                               166
Fig. A.33. Six-input EXOR gate                                167
Fig. A.34. Seven-input EXOR gate                              167
Fig. A.35. Eight-input EXOR gate                              168
Fig. A.36. NOT gate                                           168
Fig. A.37. Low-Pass Filter                                    169
Fig. A.38. Band-Pass Filter                                   169
Fig. A.39. High-Pass Filter                                   169
Fig. A.40. Clock                                              170
Fig. A.41. Keyboard Hub                                       170
Fig. A.42. Keyboard                                           170
Fig. A.43. Mach-Zehnder interferometer                        170
Fig. A.44. Matching unit                                      171
Fig. A.45. One-bit memory unit                                171
Fig. A.46. Monitor hub                                        171
Fig. A.47. Monitor                                            172
Fig. A.48. Optical amplifier                                  172
Fig. A.49. Optical coupler $1 \times 2$                       172
Fig. A.50. Optical coupler $1 \times 3$                       172
Fig. A.51. Optical coupler $1 \times 4$                       172
Fig. A.52. Optical coupler $1 \times 5$                       172
Fig. A.53. Optical coupler $1 \times 6$                       173
Fig. A.54. Optical coupler $1 \times 8$                       173
Fig. A.55. Optical coupler $1 \times 9$                       173
Fig. A.56. Optical coupler $1 \times 10$                      173
Fig. A.57. Optical coupler $1 \times 16$                      174
Fig. A.58. Optical coupler $1 \times 20$                      174
Fig. A.59. Optical coupler $1 \times 24$                      174
Fig. A.60. Optical coupler $1 \times 30$                      175
Fig. A.61. Optical coupler $2 \times 1$                       175
Fig. A.62. Optical coupler $3 \times 1$                       175
Fig. A.63. Optical coupler $4 \times 1$                       175
Fig. A.64. Optical coupler $5 \times 1$                       175
Fig. A.65. Optical coupler $6 \times 1$                       176
Fig. A.66. Optical coupler $7 \times 1$                       176
Fig. A.67. Optical coupler $8 \times 1$                       176
Fig. A.68. RAM8                                               176

Fig. A.69. RAM16    177
Fig. A.70. RAM20    177
Fig. A.71. RAM24    177
Fig. A.72. RAM30    178
Fig. A.73. ROM8    178
Fig. A.74. ROM16    178
Fig. A.75. ROM20    179
Fig. A.76. ROM24    179
Fig. A.77. ROM30    179
Fig. A.78. Optical input port    180
Fig. A.79. Optical output port    180
Fig. A.80. Optical switch    180
Fig. A.81. Spatial light modulator    181
Fig. A.82. Wavelength converter    181
Fig. A.83. Start point same-layer intermodule link    181
Fig. A.84. Start point different-layer intermodule link    181
Fig. A.85. End point same-layer intermodule link    182
Fig. A.86. End point different-layer intermodule link    182
Fig. A.87. Through-hole intermodule link    182
Fig. A.88. SR latch    182
Fig. A.89. JK latch    183
Fig. A.90. D latch    183
Fig. A.91. T latch    184
Fig. A.92. SR flip flop    184
Fig. A.93. JK flip flop    185
Fig. A.94. 5 Input JK flip flop    185
Fig. A.95. D flip flop    186
Fig. A.96. T flip flop    186

## Appendix B

Booth's Algorithm Code Fragment    188
Fig.B.1    188
Signed Divide    196
Fig.B.2    196
Floating point Add/Subtract    204
Fig.B.3    204
Floating Point Signed Multiply    216
Fig.B.4    216
Floating Point Signed Divide    231
Fig.B.5    231
Magnitude Comparator    246
Fig.B.6    246

# List of Tables

**1.10. Signed Multiplier**
Table.1.10.1. States for control logic where *ASR* stands for 'arithmetic shift right'.     23

**1.15. Logic Unit**
Table.1.15.1. Truth table for an AND gate     41
Table .1.15.2. Truth table for OR gate     42
Table .1.15.3, Truth table for EXOR gate     43
Table .1.15.4. NOT gate truth table     43

**1.16. Two-Bit Magnitude Comparator**
Table .1.16.1. Truth table for 2-bit magnitude comparator     45
Table .1.16.2     45
Table .1.16.3     46
Table .1.16.4     46
Table .1.16.5     47
Table .1.16.6     47
Table .1.16.7     48
Table .1.16.8     48
Table .1.16.9     48

**2.1.1. Instruction Set Listing for RISC-V**
Table 2.1.1.1. Instruction type frames     68
Table 2.1.1.2. RV32I instruction frames     69
Table 2.1.1.3. RV64I RV32/RV64 Zifencei RV21/RV64 Zicsr RV32M RV64M instruction type frames     69
Table 2.1.1.4. RV64i instruction frames     70
Table 2.1.1.5. RV32/RV64 Zifencei instruction frame     70
Table 2.1.1.6. RV32/64 Zicsr instruction frames     70
Table 2.1.1.7. RV32M instruction frames     71
Table 2.1.1.8. RV64M instruction frames     71
Table 2.1.1.9. RV32A RV64A instruction type frames     71
Table 2.1.1.10. RV32A instruction frames     71
Table 2.1.1.11. RV64A instruction frames     72
Table 2.1.1.12. RV32F RV64F instruction type frames     72
Table 2.1.1.13. RV32F instruction frames     73
Table 2.1.1.14. RV64F instruction frames     73
Table 2.1.1.15. RV32D RV64D instruction type frames     73
Table 2.1.1.16. RV32D instruction frames     74
Table 2.1.1.17. RV64D instruction frames     74
Table 2.1.1.18. RV32Q/RV64Q instruction type frames     75
Table 2.1.1.19. RV32Q instruction frames     75
Table 2.1.1.20. RV64Q instruction frames     76

**2.1.2. Theory and Notes**
Table.2.1.2.1. Register, width 32 bits     76
Table 2.1.2.2. Registers with width 64 bits     77

## 4.1. Holodeck Theory of Operation
## 4.1.2.8. Different Colours

| | |
|---|---|
| Table 4.1.2.8.1. Colour and wavelength table | 118 |
| Truth Table A.1 | 151 |
| Truth Table A.2 | 151 |
| Truth Table A.3 | 152 |
| Truth Table A.4 | 152 |
| Truth Table A.5 | 153 |
| Truth Table A.6 | 153 |
| Truth Table A.7 | 154 |
| Truth Table A.8 | 154 |
| Truth Table A.9 | 155 |
| Truth Table A.10 | 155 |
| Truth Table A.11 | 156 |
| Truth Table A.12 | 156 |
| Truth Table A.13 | 157 |
| Truth Table A.14 | 157 |
| Truth Table A.15 | 158 |
| Truth Table A.16 | 158 |
| Truth Table A.17 | 159 |
| Truth Table A.18 | 159 |
| Truth Table A.19 | 160 |
| Truth Table A.20 | 160 |
| Truth Table A.21 | 161 |
| Truth Table A.22 | 161 |
| Truth Table A.23 | 162 |
| Truth Table A.24 | 162 |
| Truth Table A.25 | 163 |
| Truth Table A.26 | 163 |
| Truth Table A.27 | 164 |
| Truth Table A.28 | 164 |
| Truth Table A.29 | 165 |
| Truth Table A.30 | 165 |
| Truth Table A.31 | 166 |
| Truth Table A.32 | 166 |
| Truth Table A.33 | 167 |
| Truth Table A.34 | 167 |
| Truth Table A.35 | 168 |
| Truth Table A.36 | 168 |
| Table A.37 | 182 |
| Table A.38 | 183 |
| Table A.39 | 183 |
| Table A.40 | 184 |
| Table A.41 | 184 |
| Table A.42 | 185 |
| Table A.43 | 185 |
| Table A.44 | 186 |
| Table A.45 | 186 |

# Introduction

This book is about getting a requirements together for proof of concept of the various technologies used for holodeck design.

## The book is divided into four main parts:

**Part 1.** The description of functions of a concept Photonic processor and then some optical core theory, pipelining, threading and then it shows the mappings from core to voxels and threads to force points. It also covers at a very high level how a motherboard could be put together through layering the boards. Then it covers the basics of networked and multi-user holodecks.

**Part 2.** The RISC-V instruction set and some notes. Shows mappings of hardware discussed in part 1 to RISC-V instructions.

**Part 3.** Volumetric theory with holodeck volumetrics showing the concept of breaking a voxel down into force points which are used as triangulated magnetic moments which cause reflection. It also covers a level of solidness to touch theories.

**Part 4.** This shows various holodeck set-ups, theories, and concepts.

At the end I give a conclusion and also discuss the possibility of increasing the force point count per voxel from between $1 \times 10^3$ per voxel axis to $1 \times 10^9$ per voxel axis and discuss some of the implications.

# Part 1

1.1. Introduction

**Part 1 The Theoretical Functions of a Photonic Computer**

1.1     Introduction

1.2     Optical Switches

**Processor Functions**

1.3     Shift Left Register

1.4     Shift Right Register

1.5     Arithmetic Shift Right

1.6     Circular Shift Left

1.7     Circular Shift Right

1.8     Full Adder

1.9     Adder Subtractor

1.10    Signed Multiplier

1.11    Signed Division

1.12    Floating Point Add/Subtract

1.13    Floating Point Signed Multiply

1.14    Floating Point Signed Divide

1.15    Logic Unit

1.16    Two-Bit Magnitude Comparator

**Optical Core Theory**

1.17 ALU Core Theory

Introduction to Parallelism within a Core.

1.17.1.Threading

**Holodeck Motherboard Theory**

1.18.1 Datacenters Holodeck Motherboard Rough Requirements

1.18.2 Personal Holodeck Ppv Motherboard Rough Requirements

# 1.1. Introduction

Light, when at a particular colour (wavelength), oscillates up and down like a cosine or sine wave in math (Fig. 1.1.1, Fig. 1.1.2). The light wave has two components, the electric and magnetic components which make up an electromagnetic wave.

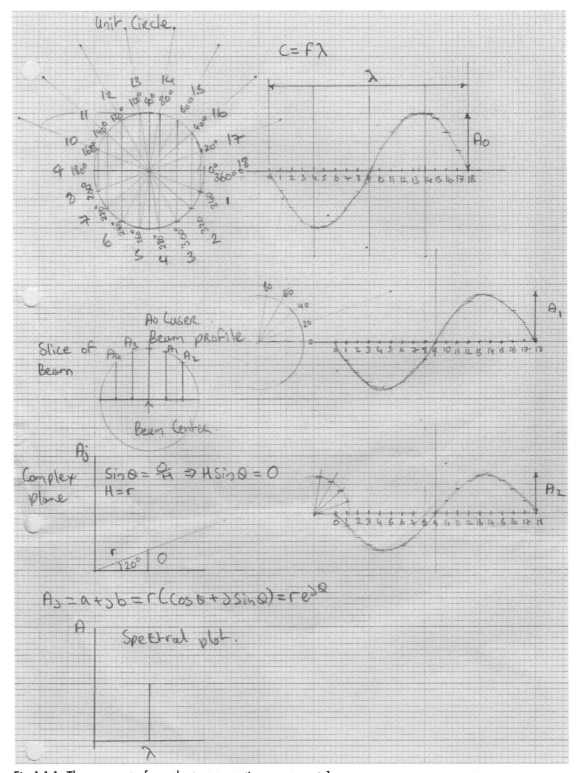

Fig.1.1.1. The concept of an electromagnetic wave, part 1

Light can be thought of as waves of a certain polarisation. For light for optical computing, the polarisation could be traverse magnetic or traverse electric.

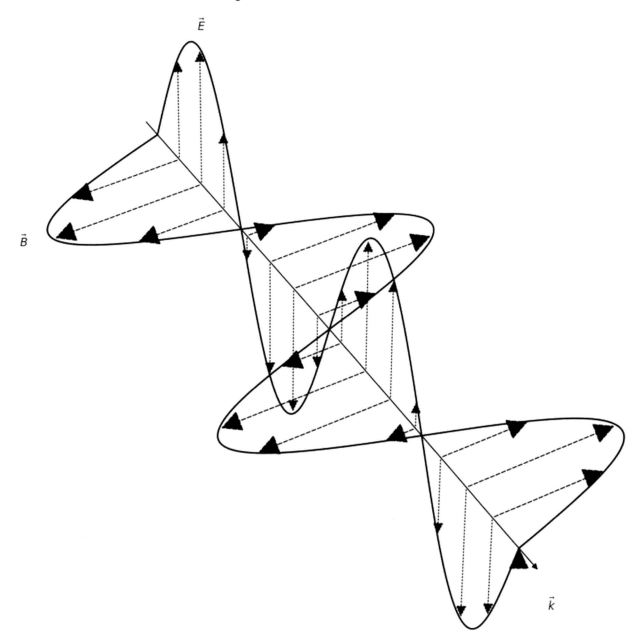

Fig.1.1.2.An electromagnetic wave

For the duration of this book, light is intensity modulated via a Mach-Zehnder interferometer. Thus the output waveform is an intensity-modulated wave where, for this book, the envelope is what is interesting. Where for a high intensity the envelope will be high and for a low intensity the envelope will show a low intensity (Fig. 1.1.3). Also the width of the information bit will be measured in wavelengths to keep the maths for matching paths simple.

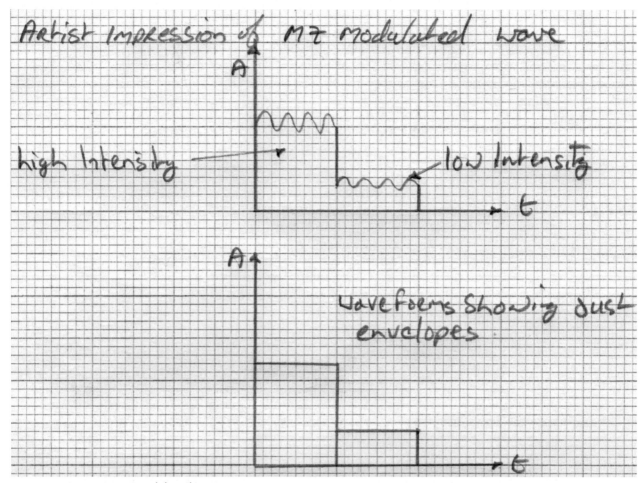

Fig.1.1.3. An intensity modulated wave

# 1.2. Optical Switches

Optical switches can be of the following forms: [31]

1. By decrease in refractive index to cause total internal reflection.
2. Change the refractive index with temperature.
3. To change the polarisation of light as it travels through the medium.
4. By using MEMS systems.
5. With the Kerr effect, either via applying an electric field or via the optical Kerr effect.
6. Via the use of liquid crystals and a polariser
7. With the use of the quantum-confined Stark effect.

# Processor Functions

In this part of the book, I am going to introduce the theory behind the functions which are inside the processor of a photonic computer. The purpose here is to introduce the reader to the processor functions in bite-sized chunks and to show them as reachable goals towards a large-scale processor design.

For clearness, in the early stages, I have shown several circuits with 4-bit or 8-bit registers to show functionality. In a real processor, I would need the registers to be 32 or 64 bits in length. For floating point examples, with the use of Java code, I use 32-bit IEEE-754-2008 standard. The reason for this is that if I used 64 bits, the examples would not fit on the page. Also note that Appendix A describes the components used in the simulations.

# 1.3. Shift Left Register

A shift left register, when clocked, shifts the data to the left.

| λ4[**0**] | λ3[**1**] | λ2[**0**] | λ1[**1**] |
|---|---|---|---|

a. Register initial set-up

| λ4[**1**] | λ3[**0**] | λ2[**1**] | λ1[**0**] |
|---|---|---|---|

b.  Register after 1 shift left

| λ4[**0**] | λ3[**1**] | λ2[**0**] | λ1[**0**] |
|---|---|---|---|

c.  Register after another shift

| λ4[**1**] | λ3[**0**] | λ2[**0**] | λ1[**0**] |
|---|---|---|---|

d.  Register after another shift

| λ4[**0**] | λ3[**0**] | λ2[**0**] | λ1[**0**] |
|---|---|---|---|

e.  Register after another shift

Fig.1.3.1.Shift left register operation

Fig. 1.3.1 shows the theory behind a left shift register. On each shift the intensity levels propagate into the next bit slot where it is at the wavelength of the bit slot shifted to. From above, λ1 is the least significand bit slot and λ4 is the most significant bit slot.

Fig. 1.3.2 shows a shift left register with its initial set-up where λ1 is the least significant bit slot and λ4 the most significant bit slot. The output ports to watch are C15, C16, C17, and C18.

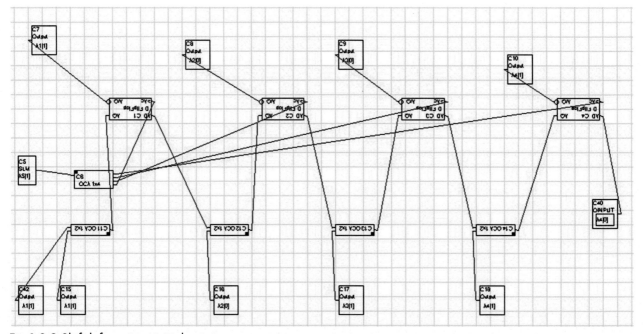

Fig.1.3.2.Shift left register initial set-up

Fig. 1.3.3 shows the shift left register with one bit shift left

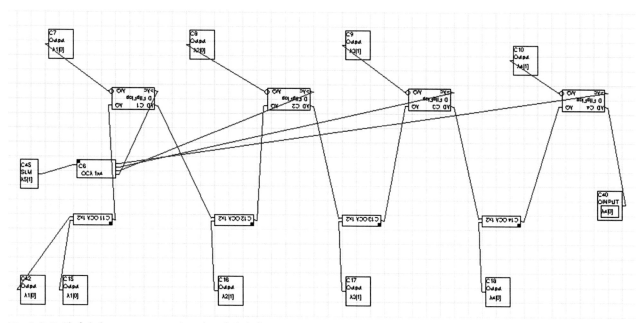

Fig.1.3.3.Shift left register one bit slot shift left

Fig. 1.3.4 shows the shift left register with two bit shifts left.

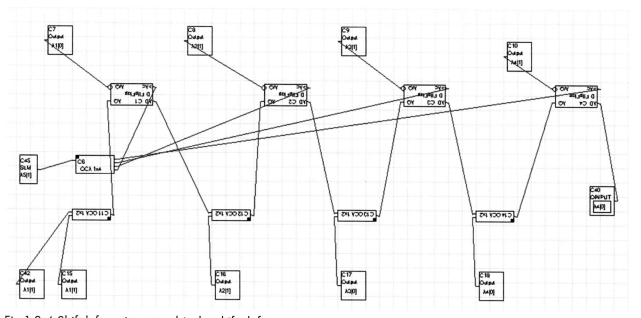

Fig.1.3.4.Shift left register two bit slot shifts left

**Michael Cloran**

Fig. 1.3.5 shows the shift left register with three bit shifts left.

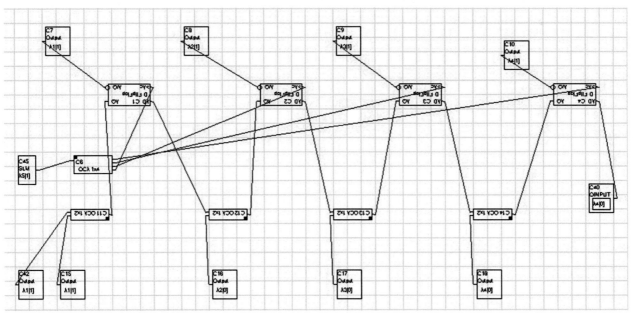

Fig.1.3.5.Shift left register three bit slot shifts left

Fig. 1.3.6 shows the shift left register with four bit shifts left

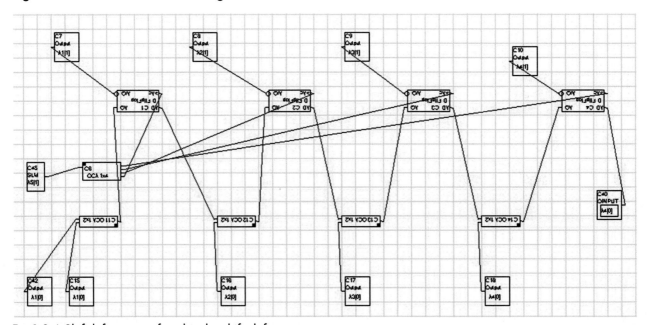

Fig.1.3.6.Shift left register four bit slot shifts left

# 1.4. Shift Right Register

A shift right register, when clocked, shifts the data to the right.

| λ4[**1**] | λ3[**0**] | λ2[**1**] | λ1[**0**] |
|---|---|---|---|

a. Register initial set-up

| λ4[**0**] | λ3[**1**] | λ2[**0**] | λ1[**1**] |
|---|---|---|---|

b. Register after a shift right

| λ4[**0**] | λ3[**0**] | λ2[**1**] | λ1[**0**] |
|---|---|---|---|

c. Register after another shift right

| λ4[**0**] | λ3[**0**] | λ2[**0**] | λ1[**1**] |
|---|---|---|---|

d. Register after another shift right

| λ4[**0**] | λ3[**0**] | λ2[**0**] | λ1[**0**] |
|---|---|---|---|

e. Register after another shift right

Fig.1.4.1 shift right register operation

Fig. 1.4.1 shows the theory behind a shift right register, where λ1 is the least significant bit slot and λ4 is the most significant bit slot. Note how the intensities propagate from one bit slot wavelength to another.

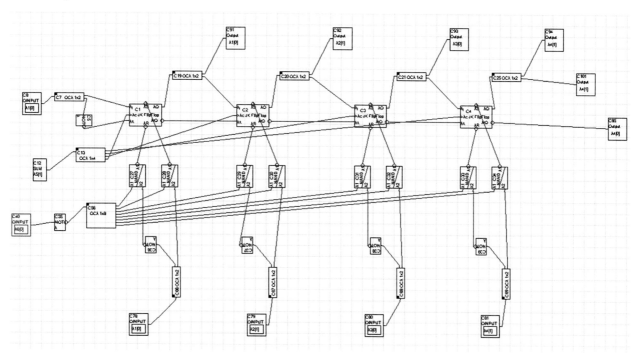

Fig.1.4.2.Optical multimode shift right register being loaded

Fig. 1.4.2 shows a multimode shift register. The modes are load values into register and shift right. When C40 optical input is at a low intensity, the values on C78, C79, C80, and C81 are loaded into the register and displayed on C91, C92, C93, and C94. It is important to note here that λ1 is the least significant bit slot; λ4 is the most significant bit slot.

Now If C40 is set to a high intensity, the shift mode of the register is used. Fig. 1.4.3 shows a shift of the register by one bit slot. C92 now has the previous value of C91. Note here that only the intensity information is shifted in that the information is translated to the next bit slot's wavelength.

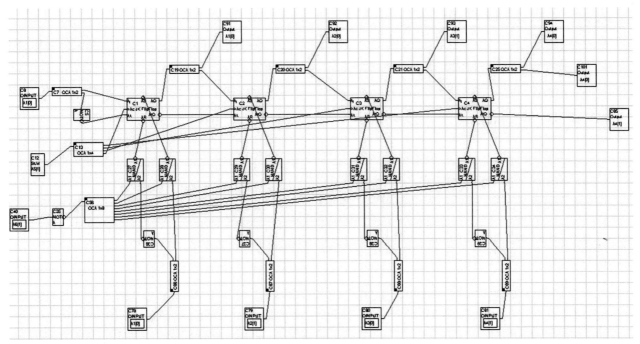

Fig.1.4.3.Optical multimode shift right register shifted by one bit slot

If C40 is left high, the shift will be to the right. Fig. 1.4.4 shows the next bit shift.

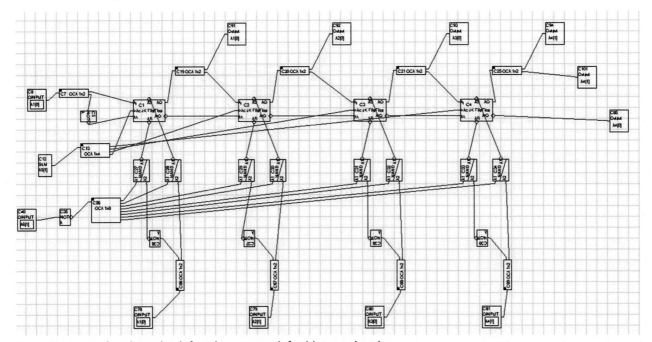

Fig.1.4.4.Optical multimode shift right register shifted by two bit slots

If C40 if still again left high, the next bit shift is implemented in Fig. 1.4.5.

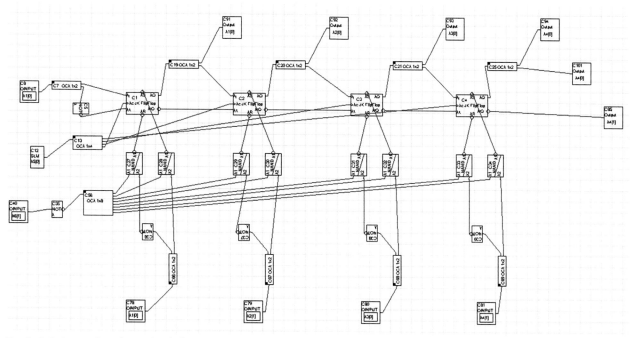

Fig.1.4.5.Optical multimode shift register shifted by three bit slots

It is important to note that the shift is a logical shift in that the value shifted into the least significant bit is a low intensity and is setup via C8.

# 1.5. Arithmetic Shift Right

Arithmetic shift right is where the sign bit is kept and propagates with the shift.

| λ4[**1**] | λ3[**0**] | λ2[**0**] | λ1[**1**] |
|-----------|-----------|-----------|-----------|

a. Initial set-up. Notice sign bit on λ4 (the most significant bit).

| λ4[**1**] | λ3[**1**] | λ2[**0**] | λ1[**0**] |
|-----------|-----------|-----------|-----------|

b. Register after a shift right.

| λ4[**1**] | λ3[**1**] | λ2[**1**] | λ1[**0**] |
|-----------|-----------|-----------|-----------|

c. Register after another shift right.

| λ4[**1**] | λ3[**1**] | λ2[**1**] | λ1[**1**] |
|-----------|-----------|-----------|-----------|

d. Register after another shift right.

| λ4[**1**] | λ3[**1**] | λ2[**1**] | λ1[**1**] |
|-----------|-----------|-----------|-----------|

e. Register after another shift right.

Fig.1.5.1.Arithmetic shift right operation

Fig. 1.5.1 shows the theory behind an arithmetic shift right register. Notice the shift right shifts the sign bit as well as the data.

# 1.6. Circular Shift Left

With a circular shift left, the most significant bit is fed back into the register at the least significant bit slot.

| λ4[**0**] | λ3[**1**] | λ2[**0**] | λ1[**1**] |
|---|---|---|---|

   a. Initial register set-up.

| λ4[**1**] | λ3[**0**] | λ2[**1**] | λ1[**0**] |
|---|---|---|---|

   b.  Register after circular shift left. Here the fourth bit of (a) is fed back into bit 1 with the shift left.

| λ4[**0**] | λ3[**1**] | λ2[**0**] | λ1[**1**] |
|---|---|---|---|

   c.  Register after another circular shift left.

| λ4[**1**] | λ3[**0**] | λ2[**1**] | λ1[**0**] |
|---|---|---|---|

   d.  Register after another circular shift left.

| λ4[**0**] | λ3[**1**] | λ2[**0**] | λ1[**1**] |
|---|---|---|---|

   e.  Register after another circular shift left.

Fig.1.6.1.Circular shift left operation

Fig. 1.6.1 shows the operation of a circular shift left register.

Fig. 1.6.2 Shows a circular shift left register with initial settings. The output ports to watch for the values when shifting are C15, C16, C17, and C18. Also λ1 is least significant bit slot, and λ4 is most significant bit slot.

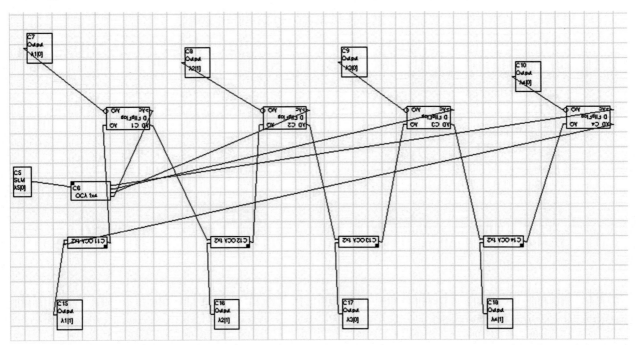

Fig. 1.6.2. Circular shift left register initial set-up

Fig. 1.6.3 shows the circular shift left register after one shift left

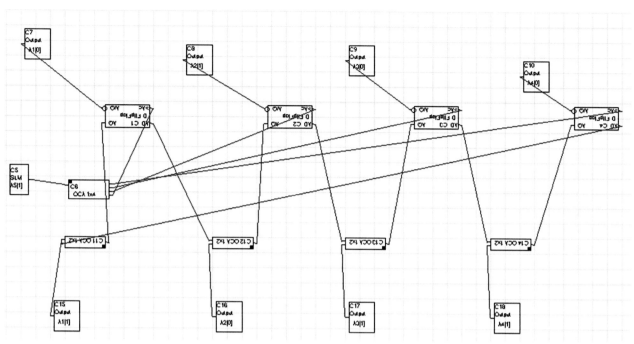

Fig. 1.6.3. Circular shift left register after one bit shift left

Fig. 1.6.4 shows the circular shift left register after another bit shift.

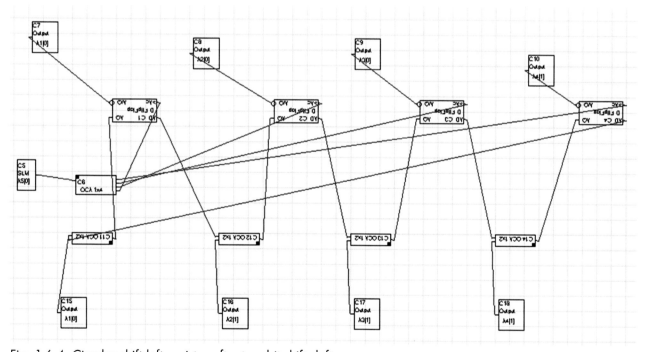

Fig. 1.6.4. Circular shift left register after two bit shifts left

Fig. 1.6.5 shows the circular shift left register after another shift.

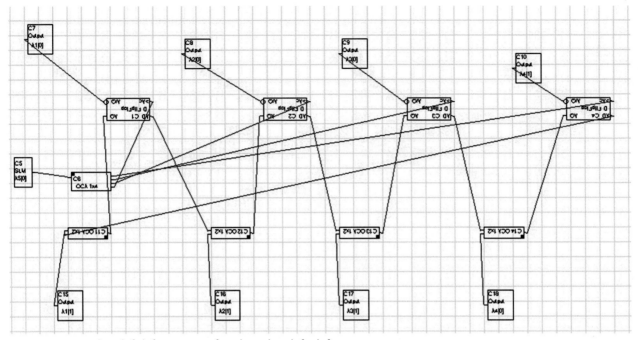

Fig. 1.6.5. Circular shift left register after three bit shifts left

Fig. 1.6.6 shows the circular shift left register after another shift left.

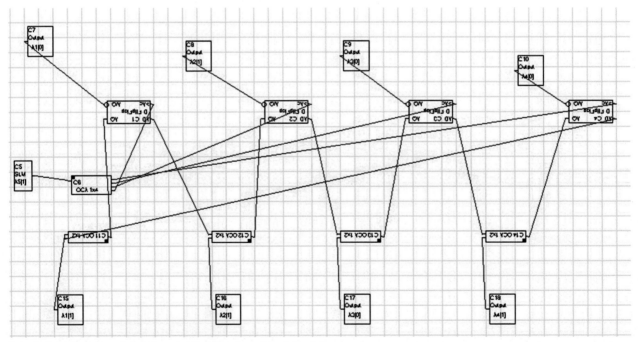

Fig. 1.6.6. Circular shift left register after four bit shifts left

# 1.7. Circular Shift Right

With a circular right shift register, the bit in the least significant bit slot is fed into the most significant bit slot on shift right.

| λ4[**0**] | λ3[**1**] | λ2[**0**] | λ1[**1**] |
|---|---|---|---|

a. Initial register set-up

| λ4[**1**] | λ3[**0**] | λ2[**1**] | λ1[**0**] |
|---|---|---|---|

b. Register after a circular shift right

| λ4[**0**] | λ3[**1**] | λ2[**0**] | λ1[**1**] |
|---|---|---|---|

c. Register after another circular right shift

| λ4[**1**] | λ3[**0**] | λ2[**1**] | λ1[**0**] |
|---|---|---|---|

d. Register after another circular shift right

| λ4[**0**] | λ3[**1**] | λ2[**0**] | λ1[**1**] |
|---|---|---|---|

e. Register after another circular shift right

Fig. 1.7.1. Circular shift right operation

Fig. 1.7.1 shows the theory behind the operation of a circular shift right register.

Fig. 1.7.2 below shows a ring counter/circular shift right register where the output of the most significant bit is fed back into the least significant bit. C12, C13, C14, and C15 show the state of the register.

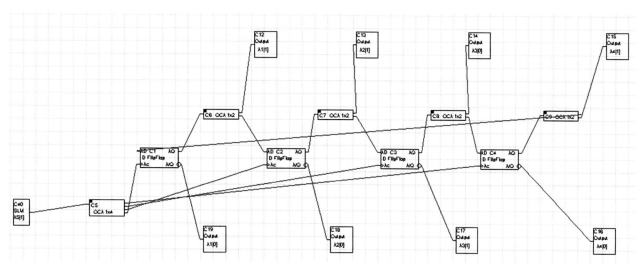

Fig. 1.7.2. Circular shift right register initial set-up

**Michael Cloran**

Fig. 1.7.3 shows the circular right shift register with one shift to the right.

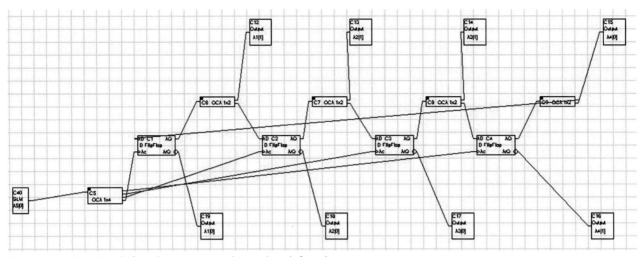

Fig. 1.7.3. Circular shift right register with one bit shift right

Fig. 1.7.4 shows the circular shift right register with another shift.

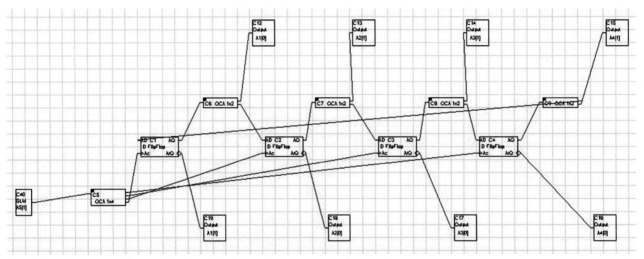

Fig. 1.7.4. Circular shift right register with two bit shifts right

Fig. 1.7.5 shows the circular shift right register with one more bit shift.

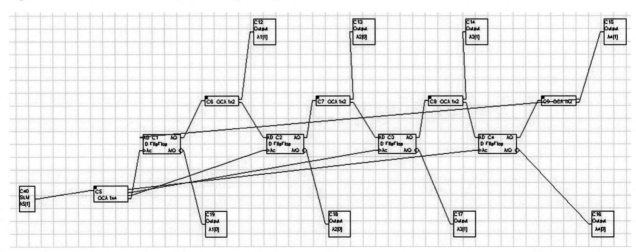

Fig. 1.7.5. Circular shift right register with three bit shifts right

16

Fig. 1.7.6 shows the circular shift right register with one more bit shift.

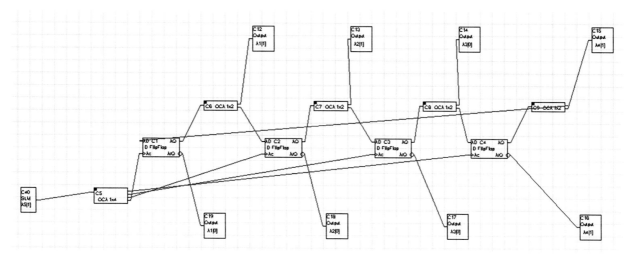

Fig. 1.7.6. Circular shift right register with four bit shifts right

# 1.8. Full Adder

A full adder adds in binary. I am going to first introduce you to a 1-bit full adder, which has three inputs and two outputs. Fig. 1.8.1 shows a 1-bit full adder circuit where Input1, Input 2, and Cin are inputs, and Sum and Cout are outputs. Before I introduce the circuit, I am going to go over basic binary addition.

```
  1
+ 0
___

  1 in binary (1 in decimal)
```

```
   1
+  1
___

 10 in binary (2 in decimal)
```

```
   1
   1
+  1
____

 11 in binary (3 in decimal)
```

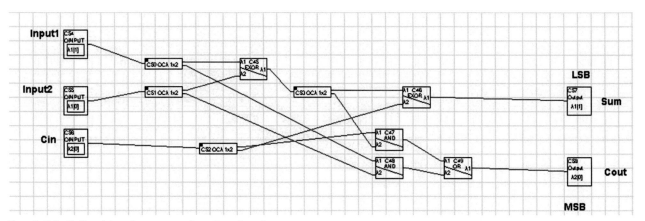

Fig. 1.8.1. Full adder doing sum 1 + 0 + 0 = 1.

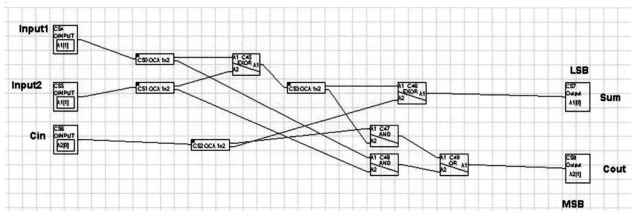

Fig. 1.8.2. Full adder showing sum 1 + 1 + 0 = 0, carry 1.

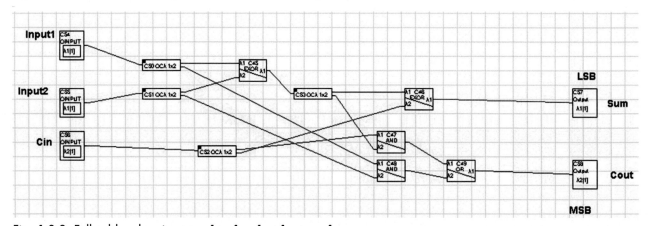

Fig. 1.8.3. Full adder showing sum 1 + 1 + 1 = 1, carry 1

I am now going to introduce a 4-bit full adder and show how the wavelengths represent the bit slots of the adder. Fig. 1.8.4 shows a 4-bit optical full adder I have marked Input1 and Input2, and the sum reads from least significant bit slot ($\lambda$1) to most significant bit slot ($\lambda$5). Fig. 1.8.4 shows the addition of 2 + 3 = 5.

Input1                              Input 2                    Sum

$$\lambda_4[\mathbf{0}]\lambda_3[\mathbf{0}]\lambda_2[\mathbf{1}]\lambda_1[\mathbf{0}] + \lambda_4[\mathbf{0}]\lambda_3[\mathbf{0}]\lambda_2[\mathbf{1}]\lambda_1[\mathbf{1}] = \lambda_4[\mathbf{0}]\lambda_3[\mathbf{1}]\lambda_2[\mathbf{0}]\lambda_1[\mathbf{1}]\ \text{decimal 5}$$

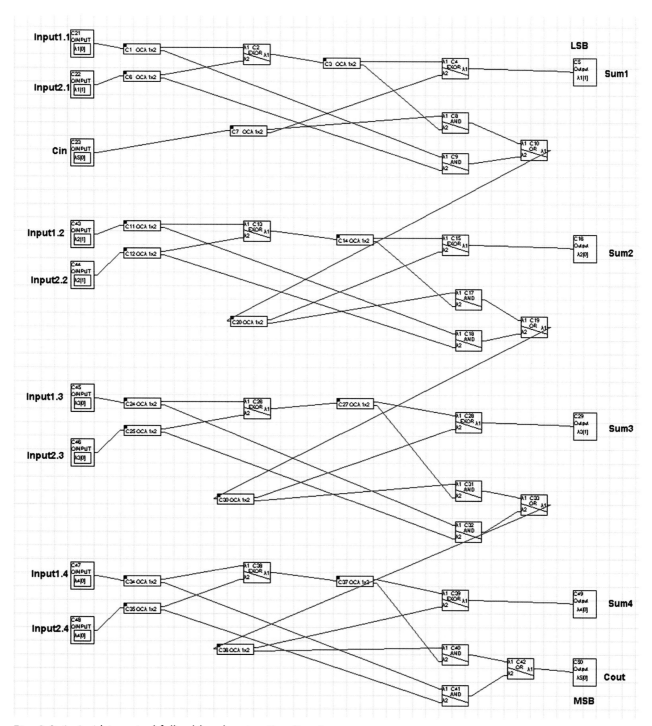

Fig. 1.8.4. A 4-bit optical full adder showing 2 + 3 = 5.

# 1.9. Adder-Subtractor

An adder-subtractor can do addition and subtraction in binary via the mode selection input, where it is at high intensity for subtraction and low intensity for addition. Before I show the circuit, I am going to explain binary subtraction and addition.

If the mode in Fig. 1.9.1 is low, a normal addition is done with Input 1 at $\lambda1[1]$ and input 2 at $\lambda2[1]$, the sum is at $\lambda1[0]$, and the carry is $\lambda2[1]$ (which is correct, as will be seen later in the 4-bit version).

Some sums in binary

$1+0 =1$

| 1 | 1 | 1 | 0001 |
| +1 | 1 | -0 | -0001 = 1110 1s complement |
| ---- | +1 | ---- | -------- + 0001 |
| 10 | -------- | 1 | ---------- |
| | 11 | | 1111 = 2's complement |

                    0001
                   +1111

                   ----------

10000 where the carry out 1 can be ignored

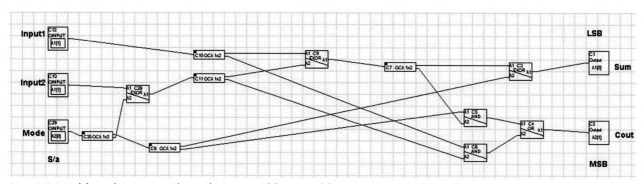

Fig. 1.9.1 Adder-subtractor with mode set to addition, adding 1 + 1 = 2 (sum = 0, carry = 1).

Now for the 1-bit adder-subtractor to do the sum 1 − 1 = 0, the mode has to be set high. In Fig. 1.9.2, the answer is again sum = $\lambda1[0]$ and the carry = $\lambda2[1]$, where the carry can be ignored for subtraction as shown above in the sums' rough work. It will make more sense in the 4-bit version.

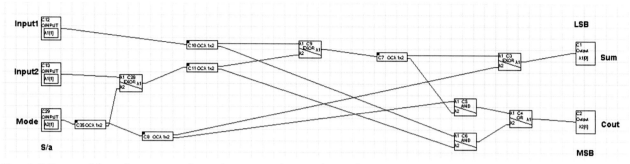

Fig. 1.9.2 an adder-subtractor with mode set to subtract doing the sum 1 – 1 = 0.

It's time now to show the 4-bit version with the same sums 1 + 1 = 2 and 1 - 1 = 0.

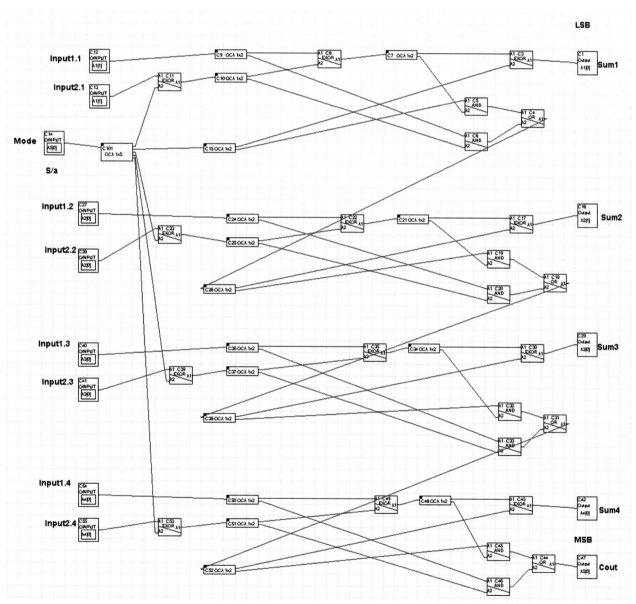

Fig. 1.9.3. The adder-subtractor in mode addition, showing a sum of 1 + 1 = 2 (10 in binary).

**Michael Cloran**

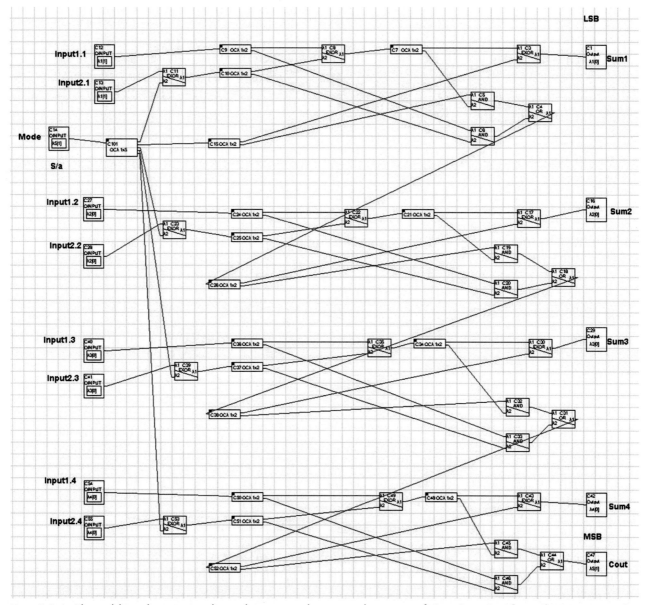

Fig. 1.9.4. The adder-subtractor with mode set to subtract, with a sum of 1 – 1 = 0, where the carry can be discarded.

# 1.10. Signed Multiplier

Optical signed multiplication can be achieved through a modified version of Booth's algorithm, where each bit slot is assigned a wavelength. Fig. 1.10.1 shows a sketch of an 8-bit register hardware design.

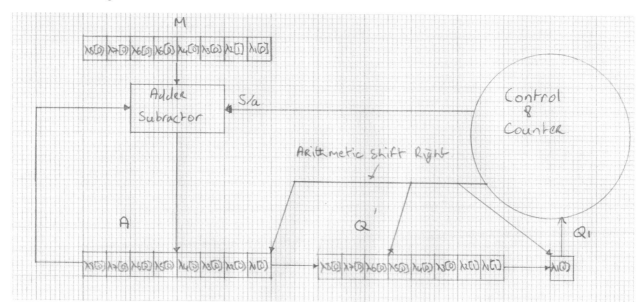

Fig. 1.10.1. Hardware showing M, A, Q, and Q1 registers initial set-up with wavelengths and intensity levels. The counter for 8-bit registers is set initially to 8.

For Booth's algorithm, the counter is set to the register size (in this case 8) and counts down to 0. The control takes Q[0] and Q1 and uses these as state variables in control logic, with the states being the following:

| Q[0] | Q1 | Action on State Event |
|---|---|---|
| $\lambda 1[0]$ | $\lambda 1[1]$ | A = A + M then ASR |
| $\lambda 1[1]$ | $\lambda 1[0]$ | A = A - M then ASR |
| $\lambda 1[1]$ | $\lambda 1[1]$ | ASR |
| $\lambda 1[0]$ | $\lambda 1[0]$ | ASR |

Table.1.10.1. States for control logic where *ASR* stands for 'arithmetic shift right'.

I have written a Java program to simulate this scenario, and it is in print in Appendix B under Booth's algorithm code fragment. The results for the above set-up is in Fig. 1.10.2. At the very end, for counter = 1. The result is shown to be 110 which is binary for 6, the correct answer for 2 × 3.

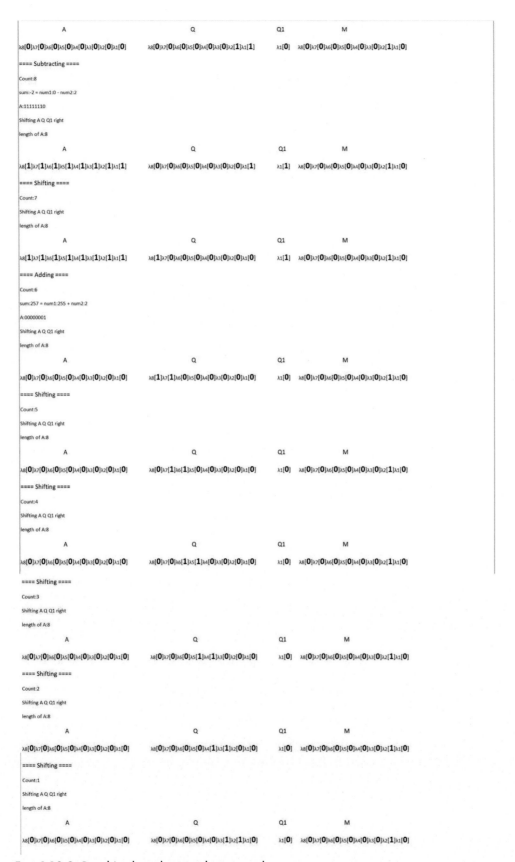

Fig. 1.10.2. Booth's algorithm simulation results

It should be said that an 8-bit multiplier unit is not what will be used but is good for a test case to show the functionality. In the real processor, 32-bit or 64-bit registers will be used.

# 1.11. Signed Division

Fig. 1.11.1 shows a block diagram of the hardware required to do signed division. Fig. 1.11.2 shows the flow chart of the control part of this diagram.

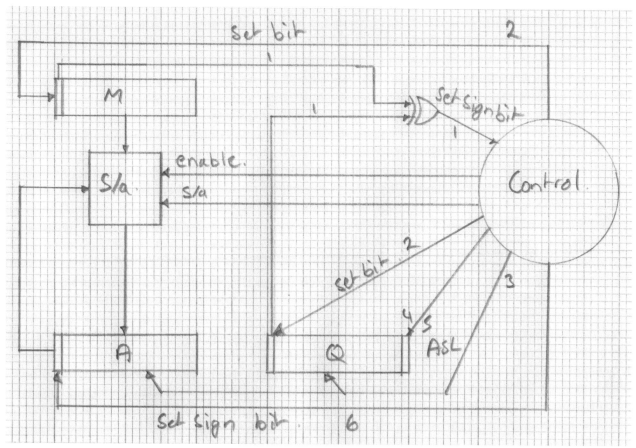

Fig. 1.11.1. Signed division concept block diagram

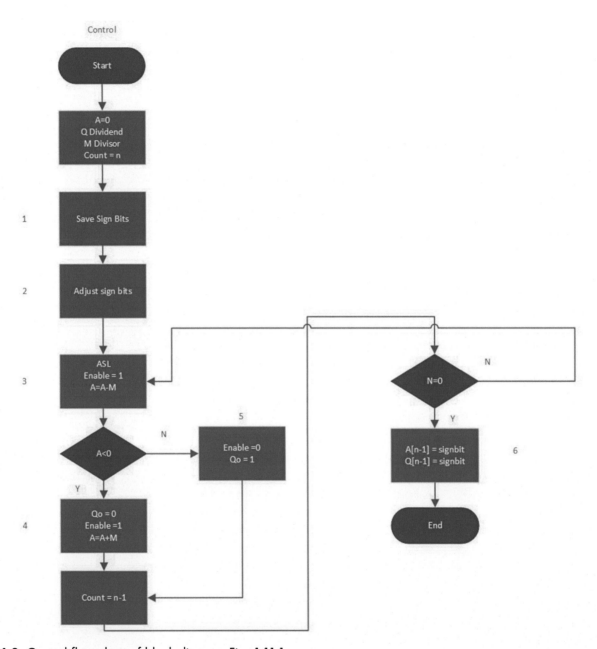

Fig. 1.11.2. Control flow chart of block diagram Fig. 1.11.1

I have written a Java program to simulate signed division and it is included in print in Appendix B. I have ran a simulation for –7 / 2 = –3, remainder –1 or 1000 0111 / 10 = 1000 0011, remainder 1000 0001 (in the example, I am using 8-bit registers). Notice the most significant bit, which is the sign bit set to 1. Below shows a run of the simulator (Fig. 1.11.3).

|  A | Q | M |
|---|---|---|
| $\lambda_8[0]\lambda_7[0]\lambda_6[0]\lambda_5[0]\lambda_4[0]\lambda_3[0]\lambda_2[0]\lambda_1[0]$ | $\lambda_8[0]\lambda_7[0]\lambda_6[0]\lambda_5[0]\lambda_4[0]\lambda_3[1]\lambda_2[1]\lambda_1[1]$ | $\lambda_8[0]\lambda_7[0]\lambda_6[0]\lambda_5[0]\lambda_4[0]\lambda_3[0]\lambda_2[1]\lambda_1[0]$ |

Count:8

==== shifting AQ left ====

Shifting A Q left

length of A:8

==== subtracting ====

sum:-2 = num1:0 - num2:2

A:11111110

==== adding ====

sum:256 = num1:254 + num2:2

A:00000000

|  A | Q | M |
|---|---|---|
| $\lambda_8[0]\lambda_7[0]\lambda_6[0]\lambda_5[0]\lambda_4[0]\lambda_3[0]\lambda_2[0]\lambda_1[0]$ | $\lambda_8[0]\lambda_7[0]\lambda_6[0]\lambda_5[0]\lambda_4[1]\lambda_3[1]\lambda_2[1]\lambda_1[0]$ | $\lambda_8[0]\lambda_7[0]\lambda_6[0]\lambda_5[0]\lambda_4[0]\lambda_3[0]\lambda_2[1]\lambda_1[0]$ |

Count:7

==== shifting AQ left ====

Shifting A Q left

length of A:8

==== subtracting ====

sum:-2 = num1:0 - num2:2

A:11111110

==== adding ====

sum:256 = num1:254 + num2:2

A:00000000

|  A | Q | M |
|---|---|---|
| $\lambda_8[0]\lambda_7[0]\lambda_6[0]\lambda_5[0]\lambda_4[0]\lambda_3[0]\lambda_2[0]\lambda_1[0]$ | $\lambda_8[0]\lambda_7[0]\lambda_6[0]\lambda_5[1]\lambda_4[1]\lambda_3[1]\lambda_2[0]\lambda_1[0]$ | $\lambda_8[0]\lambda_7[0]\lambda_6[0]\lambda_5[0]\lambda_4[0]\lambda_3[0]\lambda_2[1]\lambda_1[0]$ |

Count:6

==== shifting AQ left ====

Shifting A Q left

length of A:8

==== subtracting ====

sum:-2 = num1:0 - num2:2

A:11111110

==== adding ====

sum:256 = num1:254 + num2:2

A:00000000

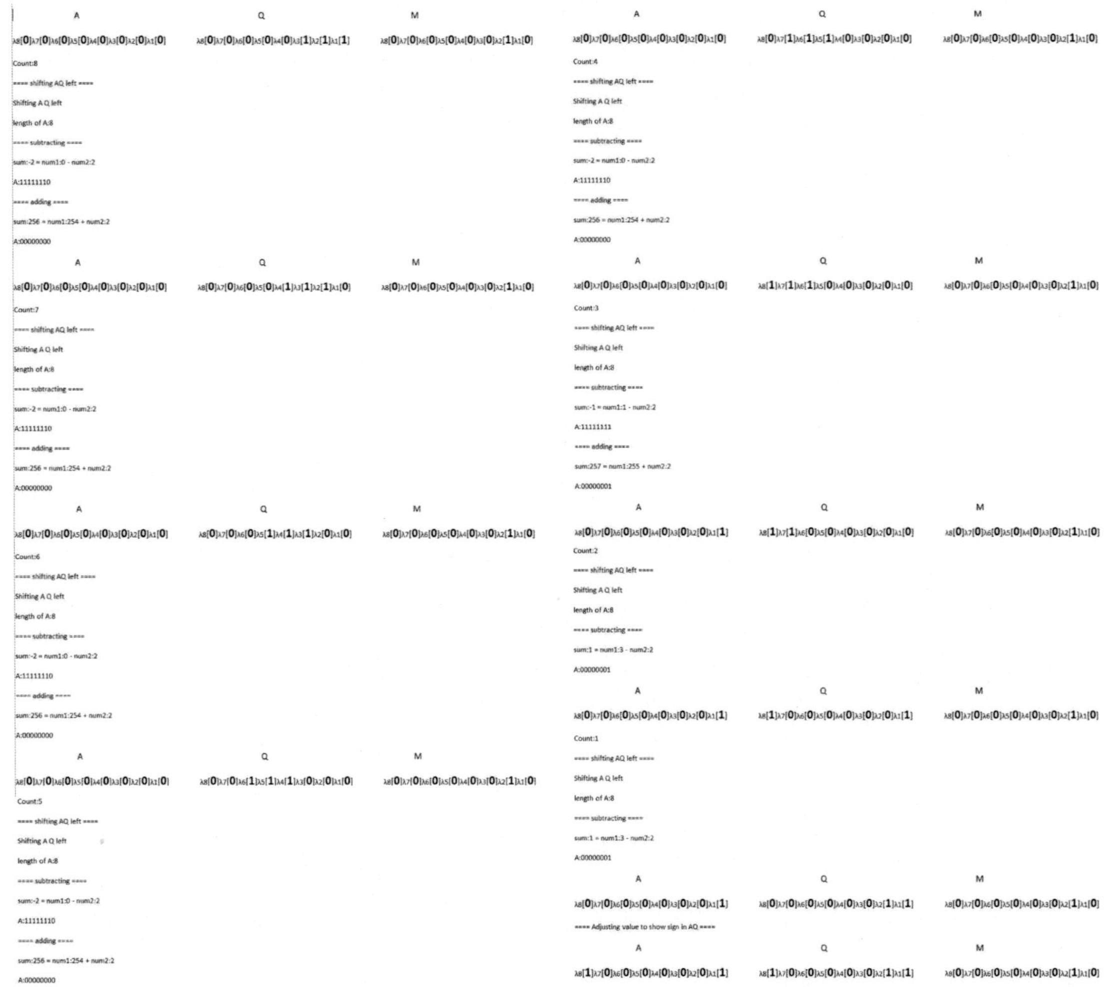

|  | A | Q | M |
|--|---|---|---|

A8[0]A7[0]A6[0]A5[0]A4[0]A3[0]A2[0]A1[0]      Q8[0]Q7[0]Q6[0]Q5[0]Q4[0]Q3[1]Q2[1]Q1[1]      M8[0]M7[0]M6[0]M5[0]M4[0]M3[0]M2[1]M1[0]

Count:8
==== shifting AQ left ====
Shifting A Q left
length of A:8
==== subtracting ====
sum:-2 = num1:0 - num2:2
A:11111110
==== adding ====
sum:256 = num1:254 + num2:2
A:00000000

A8[0]A7[0]A6[0]A5[0]A4[0]A3[0]A2[0]A1[0]      Q8[0]Q7[0]Q6[0]Q5[0]Q4[1]Q3[1]Q2[1]Q1[0]      M8[0]M7[0]M6[0]M5[0]M4[0]M3[0]M2[1]M1[0]

Count:7
==== shifting AQ left ====
Shifting A Q left
length of A:8
==== subtracting ====
sum:-2 = num1:0 - num2:2
A:11111110
==== adding ====
sum:256 = num1:254 + num2:2
A:00000000

A8[0]A7[0]A6[0]A5[0]A4[0]A3[0]A2[0]A1[0]      Q8[0]Q7[0]Q6[0]Q5[0]Q4[1]Q3[1]Q2[1]Q1[0]      M8[0]M7[0]M6[0]M5[0]M4[0]M3[0]M2[1]M1[0]

Count:6
==== shifting AQ left ====
Shifting A Q left
length of A:8
==== subtracting ====
sum:-2 = num1:0 - num2:2
A:11111110
==== adding ====
sum:256 = num1:254 + num2:2
A:00000000

A8[0]A7[0]A6[0]A5[0]A4[0]A3[0]A2[0]A1[0]      Q8[0]Q7[0]Q6[0]Q5[1]Q4[1]Q3[0]Q2[0]Q1[0]      M8[0]M7[0]M6[0]M5[0]M4[0]M3[0]M2[1]M1[0]

Count:5
==== shifting AQ left ====
Shifting A Q left
length of A:8
==== subtracting ====
sum:-2 = num1:0 - num2:2
A:11111110
==== adding ====
sum:256 = num1:254 + num2:2
A:00000000

A8[0]A7[0]A6[0]A5[0]A4[0]A3[0]A2[0]A1[0]      Q8[0]Q7[1]Q6[1]Q5[1]Q4[0]Q3[0]Q2[0]Q1[0]      M8[0]M7[0]M6[0]M5[0]M4[0]M3[0]M2[1]M1[0]

Count:4
==== shifting AQ left ====
Shifting A Q left
length of A:8
==== subtracting ====
sum:-2 = num1:0 - num2:2
A:11111110
==== adding ====
sum:256 = num1:254 + num2:2
A:00000000

A8[0]A7[0]A6[0]A5[0]A4[0]A3[0]A2[0]A1[0]      Q8[1]Q7[1]Q6[1]Q5[0]Q4[0]Q3[0]Q2[0]Q1[0]      M8[0]M7[0]M6[0]M5[0]M4[0]M3[0]M2[1]M1[0]

Count:3
==== shifting AQ left ====
Shifting A Q left
length of A:8
==== subtracting ====
sum:-1 = num1:1 - num2:2
A:11111111
==== adding ====
sum:257 = num1:255 + num2:2
A:00000001

A8[0]A7[0]A6[0]A5[0]A4[0]A3[0]A2[0]A1[1]      Q8[1]Q7[1]Q6[1]Q5[0]Q4[0]Q3[0]Q2[0]Q1[0]      M8[0]M7[0]M6[0]M5[0]M4[0]M3[0]M2[1]M1[0]

Count:2
==== shifting AQ left ====
Shifting A Q left
length of A:8
==== subtracting ====
sum:1 = num1:3 - num2:2
A:00000001

A8[0]A7[0]A6[0]A5[0]A4[0]A3[0]A2[0]A1[1]      Q8[1]Q7[0]Q6[0]Q5[0]Q4[0]Q3[0]Q2[0]Q1[1]      M8[0]M7[0]M6[0]M5[0]M4[0]M3[0]M2[1]M1[0]

Count:1
==== shifting AQ left ====
Shifting A Q left
length of A:8
==== subtracting ====
sum:1 = num1:3 - num2:2
A:00000001

A8[0]A7[0]A6[0]A5[0]A4[0]A3[0]A2[0]A1[1]      Q8[0]Q7[0]Q6[0]Q5[0]Q4[0]Q3[0]Q2[1]Q1[1]      M8[0]M7[0]M6[0]M5[0]M4[0]M3[0]M2[1]M1[0]

==== Adjusting value to show sign in AQ ====

A8[1]A7[0]A6[0]A5[0]A4[0]A3[0]A2[0]A1[1]      Q8[1]Q7[0]Q6[0]Q5[0]Q4[0]Q3[0]Q2[1]Q1[1]      M8[0]M7[0]M6[0]M5[0]M4[0]M3[0]M2[1]M1[0]

Fig. 1.11.3. Simulation results for signed division for –7 / 2 = –3, remainder –1

# 1.12. Floating Point Add/Subtract

For floating point addition and subtraction, the process is broken down into four steps.[8]

1. Check for zeros.
2. Align the significands.
3. Add or subtract the significands.
4. Normalise the result.

I have broken down the flow chart of the original[8] into manageable chunks, each of which are shown below.

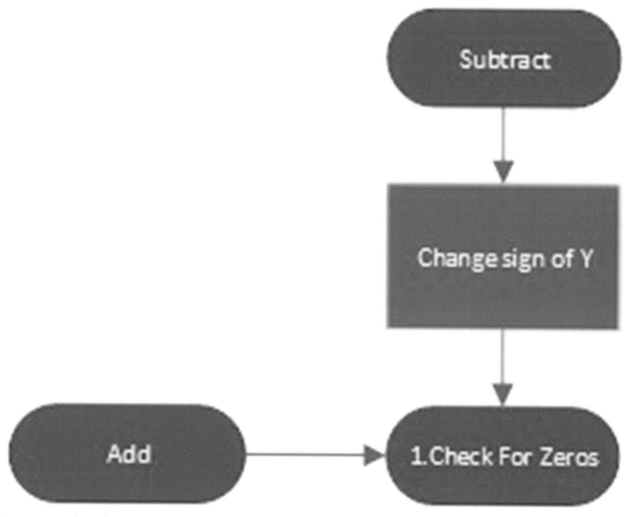

Fig. 1.12.1. Flow chart

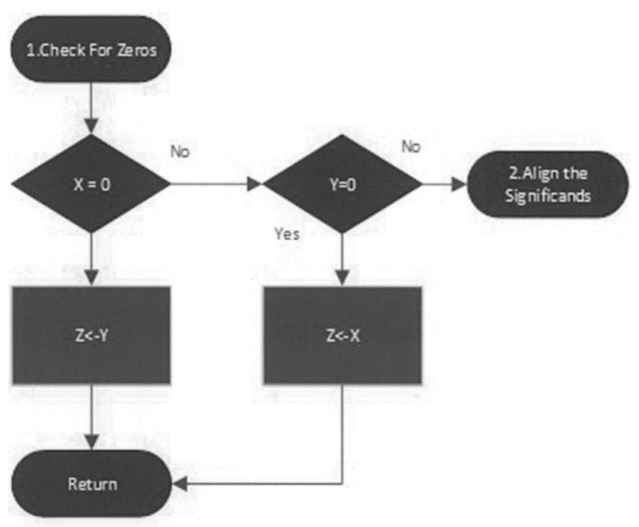

Fig. 1.12.2. Flow chart

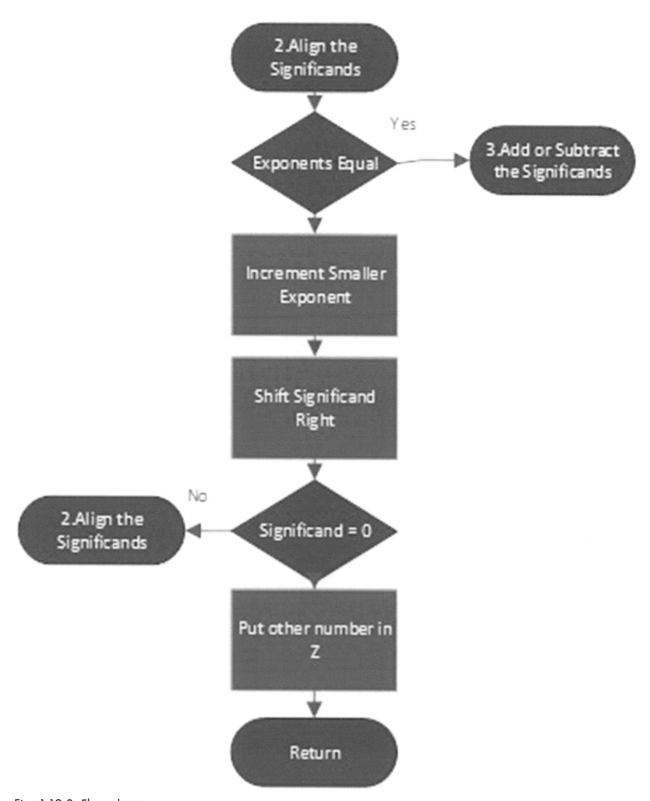

Fig. 1.12.3. Flow chart

**Michael Cloran**

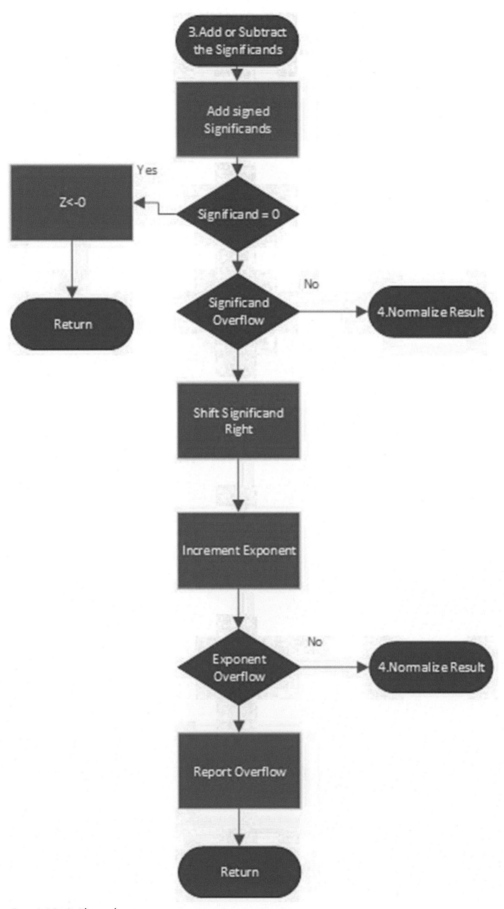

Fig. 1.12.4. Flow chart

32

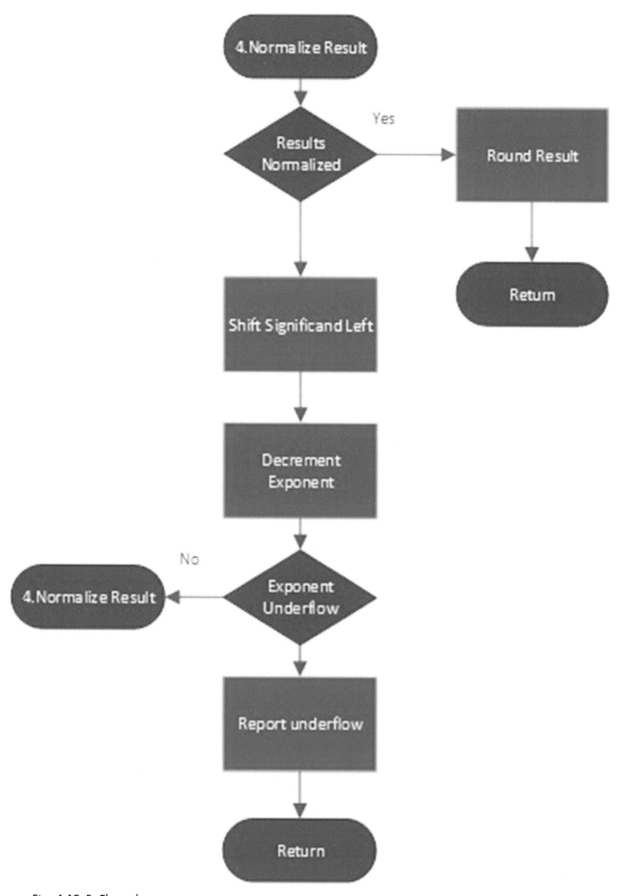

Fig. 1.12.5. Flow chart

**Michael Cloran**

I have also written a Java program to simulate floating point addition and subtraction, which is included in print in Appendix B. I have run a simulation floating point 32-bit single-format IEEE-754 for

MSB Sign Exponent    Significand                LSB

X =    0, 0000 0011, 000 0000 0000 0000 0000 0010

Y =    0, 0000 0010, 000 0000 0000 0000 0000 0111

I have left out the wavelengths here for clarity, where λ1 is the least significant bit slot and λ32 is the most significant bit slot.

And setting signbitY = 1, signbitZ = signbitY in the java program for subtraction.

The results from the program show

Aligning significands

 Z:–12 A:2 B:14

Normalising results

In the Java code I do not implement the rounding of the result.

InputX  λ32[0]      λ31[0]λ30[0]λ29[0]λ28[0]λ27[0]λ26[0]λ25[1]λ24[1]
        λ23[0]λ22[0]λ21[0]λ20[0]λ19[0]λ18[0]λ17[0]λ16[0]λ15[0]λ14[0]λ13[0]
      λ12[0]λ11[0]λ10[0] λ9[0]λ8[0]λ7[0]λ6[0]λ5[0]λ4[0]λ3[0]λ2[1]λ1[0]
InputY  λ32[1]      λ31[0]λ30[0]λ29[0]λ28[0]λ27[0]λ26[0]λ25[1]λ24[1]
        λ23[0]λ22[0]λ21[0]λ20[0]λ19[0]λ18[0]λ17[0]λ16[0]λ15[0]λ14[0]λ13[0]λ12[0]
      λ11[0]λ10[0] λ9[0]λ8[0]λ7[0]λ6[0]λ5[0]λ4[1]λ3[1]λ2[1]λ1[0]
ResultZ  λ32[1]      λ31[0]λ30[0]λ29[0]λ28[0]λ27[0]λ26[0]λ25[1]λ24[1]
        λ23[0]λ22[0]λ21[0]λ20[0]λ19[0]λ18[0]λ17[0]λ16[0]λ15[0]λ14[0]λ13[0]λ12[0]λ11[
      0]λ10[0] λ9[0]λ8[0]λ7[0]λ6[0]λ5[0]λ4[1]λ3[1]λ2[0]λ1[0]

I ran another simulation with the same values only for addition, and the results are as follows:

Aligning significands

 Z:16 A:2 B:14

Normalising results



Normalising results



Normalising results

Exponent underflow

InputX       ʎ32[0]ʎ31[0]ʎ30[0]ʎ29[0]ʎ28[0]ʎ27[0]ʎ26[0]ʎ25[1]ʎ24[1]
                  ʎ23[0]ʎ22[0]ʎ21[0]ʎ20[0]ʎ19[0]ʎ18[0]ʎ17[0]ʎ16[0]ʎ15[0]ʎ14[0]ʎ13[
        0] ʎ12[0]ʎ11[0]ʎ10[0] ʎ9[0]ʎ8[0]ʎ7[0]ʎ6[0]ʎ5[0]ʎ4[0]ʎ3[0]ʎ2[1]ʎ1[0]

InputY       ʎ32[0]ʎ31[0]ʎ30[0]ʎ29[0]ʎ28[0]ʎ27[0]ʎ26[0]ʎ25[1]ʎ24[1]
                  ʎ23[0]ʎ22[0]ʎ21[0]ʎ20[0]ʎ19[0]ʎ18[0]ʎ17[0]ʎ16[0]ʎ15[0]ʎ14[0]ʎ13[
        0] ʎ12[0]ʎ11[0]ʎ10[0] ʎ9[0]ʎ8[0]ʎ7[0]ʎ6[0]ʎ5[0]ʎ4[1]ʎ3[1]ʎ2[1]ʎ1[0]

ResultZ      ʎ32[0]ʎ31[0]ʎ30[0]ʎ29[0]ʎ28[0]ʎ27[0]ʎ26[0]ʎ25[0]ʎ24[0]
                  ʎ23[0]ʎ22[0]ʎ21[0]ʎ20[0]ʎ19[0]ʎ18[0]ʎ17[0]ʎ16[0]ʎ15[0]ʎ14[0]ʎ13[
        0] ʎ12[0]ʎ11[0]ʎ10[0] ʎ9[0]ʎ8[1]ʎ7[0]ʎ6[0]ʎ5[0]ʎ4[0]ʎ3[0]ʎ2[0]ʎ1[0]

This shows the result shifted by 3.

# 1.13. Floating Point Signed Multiply

Floating point signed multiply builds on signed multiply. Also, for floating point signed multiply for IEEE-754, the exponents are stored in biased form, which means for 32-bit floating point numbers they have 1 sign bit, 8-bit exponent, and 23-bit significand. The exponent is in biased form, which means you subtract 127 from it to get the unbiased form.

Below shows the flow charts for the floating point multiply algorithm.[8]

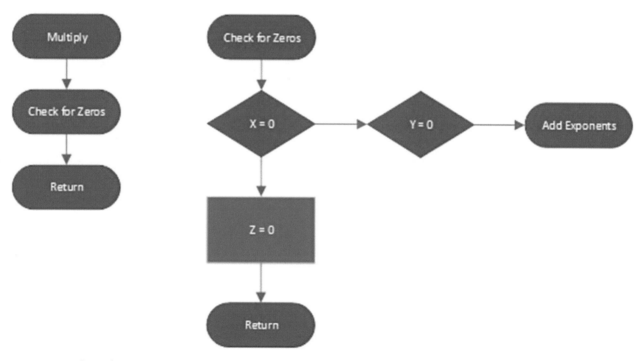

Fig. 1.13.1. Flow chart

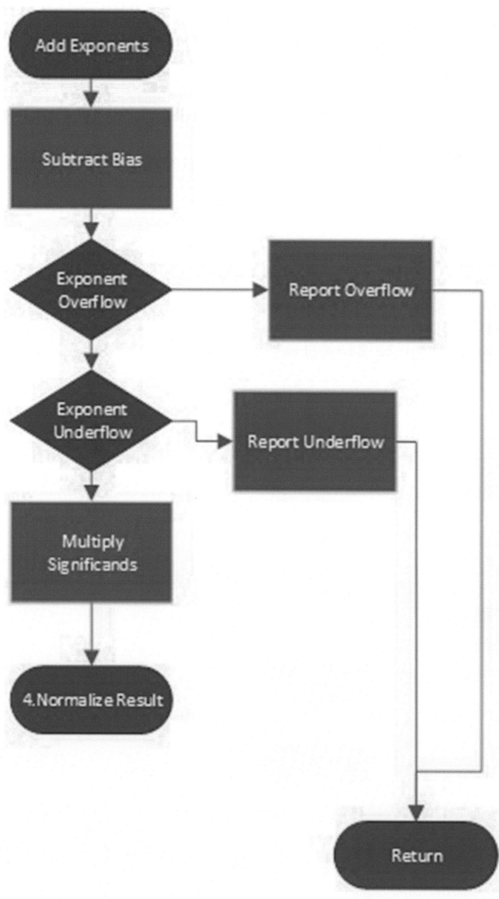

Fig. 1.13.2. Flow chart

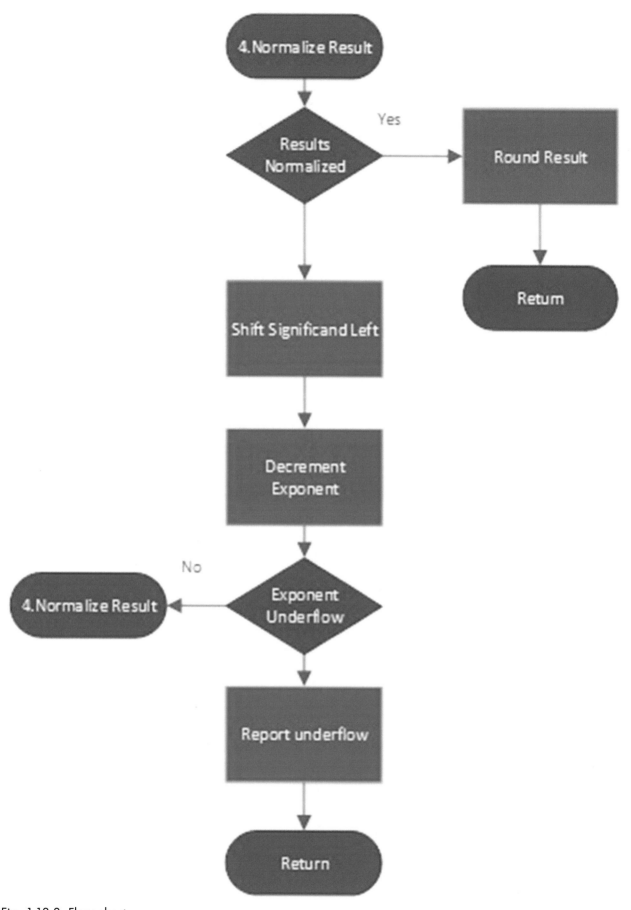

Fig. 1.13.3. Flow chart

With the flow charts as guides, I have written a Java program to simulate the workings of the floating point signed multiply, which is included in print in Appendix B.

The results of a simulation are shown below:

InputX λ32[1]　　λ31[1]λ30[0]λ29[0]λ28[0]λ27[0]λ26[0]λ25[0]λ24[0]
　　　　λ23[0]λ22[0]λ21[0]λ20[0]λ19[0]λ18[0]λ17[0]λ16[0]λ15[0]λ14[0]λ13[0]λ12[0]λ11[0]λ10[0]λ9[0]
λ8[0]λ7[0]λ6[0]λ5[0]λ4[0]λ3[0]λ2[1]λ1[0]biased form

InputY λ32[0]　　λ31[0]λ30[0]λ29[0]λ28[0]λ27[0]λ26[0]λ25[1]λ24[1]
　　　　λ23[0]λ22[0]λ21[0]λ20[0]λ19[0]λ18[0]λ17[0]λ16[0]λ15[0]λ14[0]λ13[0]λ12[0]λ11[0]λ10[0]λ9[0]
λ8[0]λ7[0]λ6[0]λ5[0]λ4[0]λ3[1]λ2[1]λ1[1]biased form

ResultZ　　　　λ32[1]　　λ31[0]λ30[0]λ29[0]λ28[0]λ27[0]λ26[1]λ25[0]λ24[0]
　　　　λ23[0]λ22[0]λ21[0]λ20[0]λ19[0]λ18[0]λ17[0]λ16[0]λ15[0]λ14[0]λ13[0]λ12[0]λ11[0]
λ10[0]λ9[0] λ8[0]λ7[0]λ6[0]λ5[0]λ4[1]λ3[1]λ2[1]λ1[0]unbiased form

# 1.14. Floating Point Signed Divide

The floating point signed divide builds on signed divide. The flow chart is shown below.[8] I have implemented Java code to simulate the flow chart, and it is useful to generate possible results. The Java code is in Appendix B in print in this book.

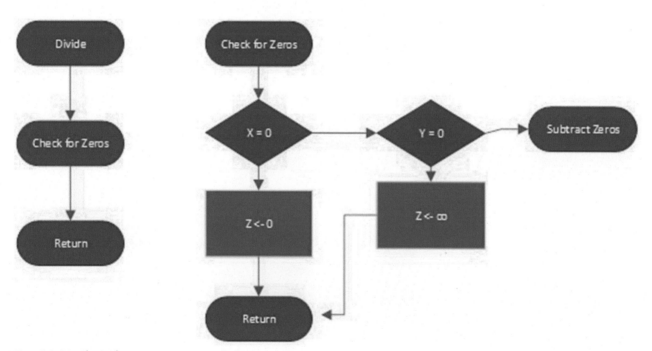

Fig. 1.14.1. Flow chart

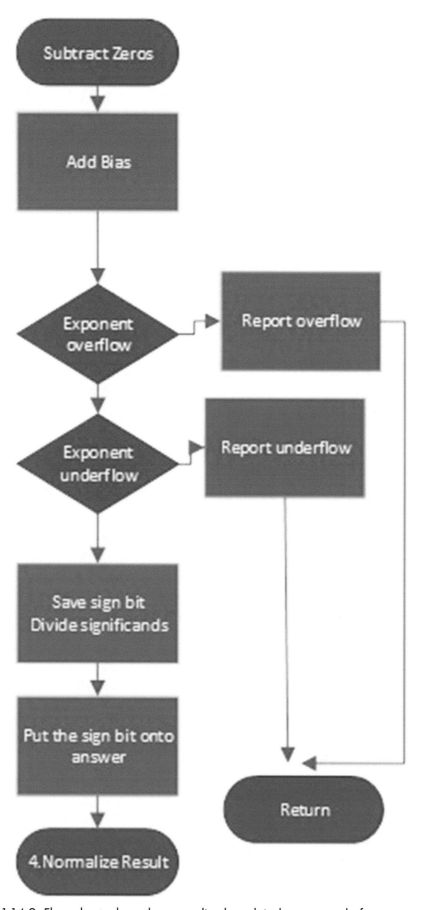

Fig. 1.14.2. Flow chart where the normalised result is the same as before.

**Michael Cloran**

With the code in Appendix B for floating point signed divide, I ran a simulation with the following initial set-up:

A = 000 0000 0000 0000 0000 0000 $\times\ 2^{0000\ 0000}$

Q = 000 0000 0000 0000 0000 0111 $\times\ 2^{1000\ 0001}$

M = 000 0000 0000 0000 0000 0011 $\times\ 2^{0000\ 0011}$

I have included in the algorithm the IEEE-754 bias, and in the results, I show InputX and InputY in biased form and ResultZ in unbiased form.

The results are below.

InputX   ʌ32[0]   ʌ31[1]ʌ30[0]ʌ29[0]ʌ28[0]ʌ27[0]ʌ26[0]ʌ25[0]ʌ24[1]
ʌ23[0]ʌ22[0]ʌ21[0]ʌ20[0]ʌ19[0]ʌ18[0]ʌ17[0]ʌ16[0]ʌ15[0]ʌ14[0]ʌ13[0]
ʌ12[0]ʌ11[0]ʌ10[0] ʌ9[0]ʌ8[0]ʌ7[0]ʌ6[0]ʌ5[0]ʌ4[0]ʌ3[1]ʌ2[1]ʌ1[1]

Biased form

InputY   ʌ32[0]   ʌ31[0]ʌ30[0]ʌ29[0]ʌ28[0]ʌ27[0]ʌ26[0]ʌ25[1]ʌ24[1]
ʌ23[0]ʌ22[0]ʌ21[0]ʌ20[0]ʌ19[0]ʌ18[0]ʌ17[0]ʌ16[0]ʌ15[0]ʌ14[0]ʌ13[0]
ʌ12[0]ʌ11[0]ʌ10[0] ʌ9[0]ʌ8[0]ʌ7[0]ʌ6[0]ʌ5[0]ʌ4[0]ʌ3[0]ʌ2[1]ʌ1[1]

Biased form

ResultZ   ʌ32[0]   ʌ31[1]ʌ30[1]ʌ29[1]ʌ28[1]ʌ27[1]ʌ26[1]ʌ25[0]ʌ24[1]
ʌ23[0]ʌ22[0]ʌ21[0]ʌ20[0]ʌ19[0]ʌ18[0]ʌ17[0]ʌ16[0]ʌ15[0]ʌ14[0]ʌ13[0]
ʌ12[0]ʌ11[0]ʌ10[0] ʌ9[0]ʌ8[0]ʌ7[0]ʌ6[0]ʌ5[0]ʌ4[0]ʌ3[0]ʌ2[1]ʌ1[0]

Unbiased form

# 1.15. Logic Unit

In this section of the book, logic arrays will be discussed by showing a simulation of the logic with a screen grab of the circuit. I am going to introduce AND, OR, and EXOR, and I am also going to cover a NOT gate array.

A logic functional block does the logic, which adheres to a rule set for that particular logic function. This rule set is defined in what is called a truth table for the logic function.

### AND Array

Table.1.15.1. Truth table for an AND gate

| A | B | Output |
|---|---|---|
| $\lambda_1[0]$ | $\lambda_2[0]$ | $\lambda_1[0]$ |
| $\lambda_1[1]$ | $\lambda_2[0]$ | $\lambda_1[0]$ |
| $\lambda_1[0]$ | $\lambda_2[1]$ | $\lambda_1[0]$ |
| $\lambda_1[1]$ | $\lambda_2[1]$ | $\lambda_1[1]$ |

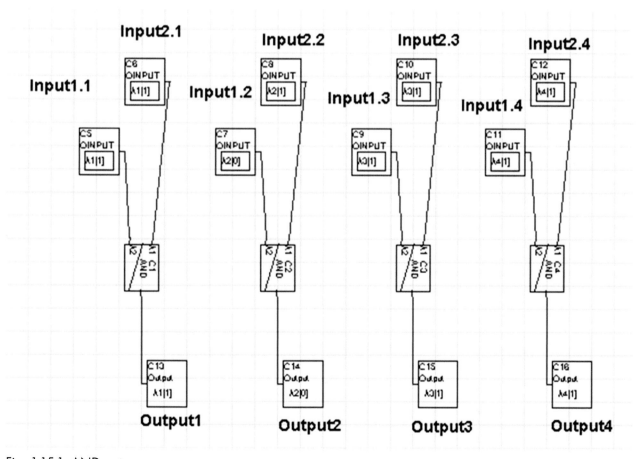

Fig. 1.15.1. AND gate array

Fig. 1.15.1 shows an AND gate array with Input1 λ1[1]λ2[0]λ3[1]λ4[1] and Input2 λ1[1]λ2[1]λ3[1]λ4[1], and the output after the AND logic is processed on each bit slot λ1[1]λ2[0]λ3[1]λ4[1]. It can be clearly seen from the truth table that for Input1.1 AND Input2.1 = Output1, which is λ1[1].

## OR Array

Table .1.15.2. Truth table for OR gate

| A | B | Output |
|---|---|--------|
| $\lambda_1[0]$ | $\lambda_2[0]$ | $\lambda_1[0]$ |
| $\lambda_1[1]$ | $\lambda_2[0]$ | $\lambda_1[1]$ |
| $\lambda_1[0]$ | $\lambda_2[1]$ | $\lambda_1[1]$ |
| $\lambda_1[1]$ | $\lambda_2[1]$ | $\lambda_1[1]$ |

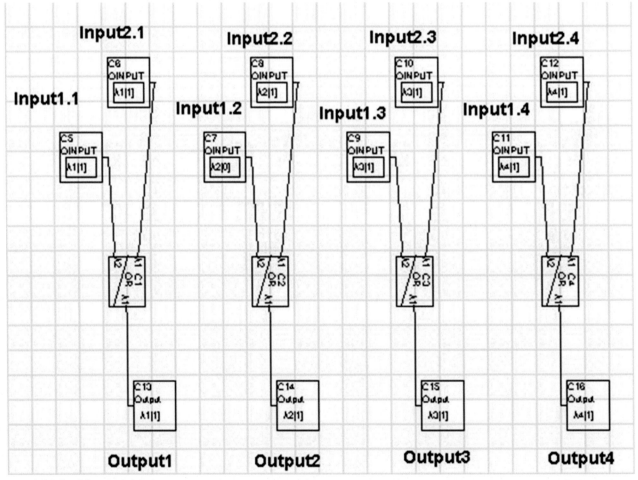

Fig. 1.15.2. OR gate array

Fig. 1.15.2 shows an OR gate array where, from the truth table, it can be clearly seen that the Input1.2 AND Input2.2 = Output2, which is λ2[1].

## EXOR Array

Table .1.15.3, Truth table for EXOR gate

| A | B | Output |
|---|---|---|
| $\lambda_1[0]$ | $\lambda_2[0]$ | $\lambda_1[0]$ |
| $\lambda_1[1]$ | $\lambda_2[0]$ | $\lambda_1[1]$ |
| $\lambda_1[0]$ | $\lambda_2[1]$ | $\lambda_1[1]$ |
| $\lambda_1[1]$ | $\lambda_2[1]$ | $\lambda_1[0]$ |

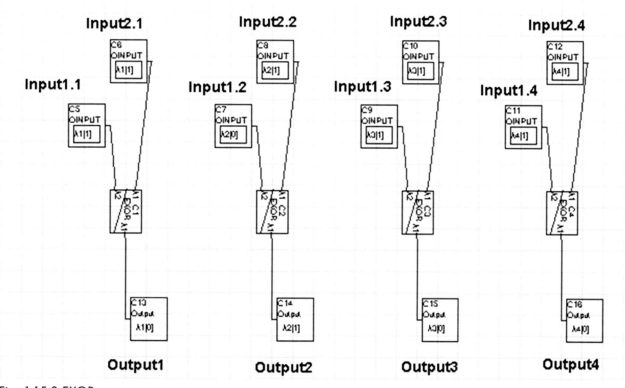

Fig. 1.15.3 EXOR gate array

Fig. 1.15.3 shows an EXOR gate array where most of the outputs are at a low intensity per the truth table, but Input1.2 EXOR Input2.2 = Output2, which is $\lambda_2[1]$ per the truth table.

## NOT Array

Table .1.15.4. NOT gate truth table

| A | Output |
|---|---|
| $\lambda_1[0]$ | $\lambda_1[1]$ |
| $\lambda_1[1]$ | $\lambda_1[0]$ |

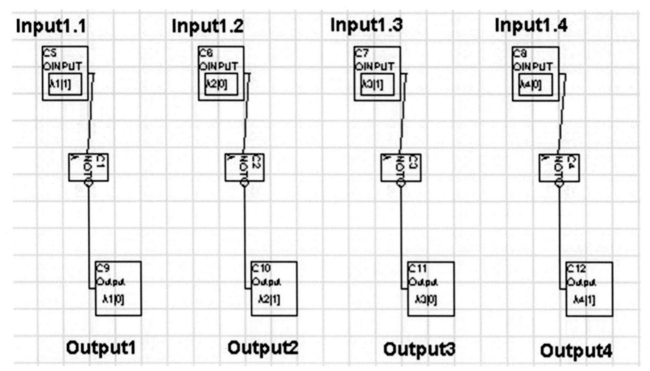

Fig. 1.15.4 NOT gate array

Fig. 1.15.4 shows a NOT gate array where if the input is at high intensity ([1]), the output is at low intensity ([0]); if the input is at low intensity, the output is at high intensity. This can be clearly seen in the diagram above.

# 1.16. Two-Bit Magnitude Comparator

A magnitude comparator is used to compare values—for instance, if a variable is less than another or if the variables are equal or if one is greater than the other. The flags in the status register can be used to hold the magnitude comparator results. Via the use of this circuit and the status register, it is possible to do branch-on-condition jumps to routines.

For simplicity, I am going to introduce the magnitude comparator in simplistic form, as in the real computer this would have to compare values for 32-bit or 64-bit registers.

The number of variables for a 2-bit magnitude comparator is 4, and $2^4 = 16$. So there are 16 combinations for the truth table T.1.16.1. A typical truth table for a 2-bit magnitude comparator is shown below for A < B, A = B, and A > B.

| Decimal | A2A1 | B2B1 | A < B | A = B | A > B |
|---------|------|------|-------|-------|-------|
| 0 | 00 | 00 | 0 | 1 | 0 |
| 1 | 00 | 01 | 1 | 0 | 0 |
| 2 | 00 | 10 | 1 | 0 | 0 |
| 3 | 00 | 11 | 1 | 0 | 0 |
| 4 | 01 | 00 | 0 | 0 | 1 |
| 5 | 01 | 01 | 0 | 1 | 0 |

| Decimal | A2A1 | B2B1 | A < B | A = B | A > B |
|---------|------|------|-------|-------|-------|
| 6 | 01 | 10 | 1 | 0 | 0 |
| 7 | 01 | 11 | 1 | 0 | 0 |
| 8 | 10 | 00 | 0 | 0 | 1 |
| 9 | 10 | 01 | 0 | 0 | 1 |
| 10 | 10 | 10 | 0 | 1 | 0 |
| 11 | 10 | 11 | 1 | 0 | 0 |
| 12 | 11 | 00 | 0 | 0 | 1 |
| 13 | 11 | 01 | 0 | 0 | 1 |
| 14 | 11 | 10 | 0 | 0 | 1 |
| 15 | 11 | 11 | 0 | 1 | 0 |

Table .1.16.1. Truth table for 2-bit magnitude comparator

I am going to use the tabular method to solve for the circuit. I have included in Appendix B some Java code to generate the truth table and to solve for one of the instances shown in the truth table A < B, A = B, and A > B. I am first going to solve for A < B using the tabular method.

From the magnitudeComparator2 Java run, it generates this table:

| | |
|---|---|
| *GroupNumber:1* | |
| mt:1 binaryRep:1000 | TickedBool: |
| *GroupNumber:1* | |
| mt:2 binaryRep:0100 | TickedBool: |
| *GroupNumber:2* | |
| mt:3 binaryRep:1100 | TickedBool: |
| *GroupNumber:2* | |
| mt:6 binaryRep:0110 | TickedBool: |
| *GroupNumber:3* | |
| mt:7 binaryRep:1110 | TickedBool: |
| *GroupNumber:3* | |
| mt:11 binaryRep:1101 | TickedBool: |

Table .1.16.2

This table groups the variable A2A1B2B1 by the number of 1s in its binary representation of the decimal value. For example, in group number 3, midterm 7 (which is 0111 in binary) has three 1s in it, thus it belongs in group number 3.

45

This solver then groups the nth group with the nth + 1 group where the row difference is only a single 1 in the binary value of the decimal midterm. This gives the following table:

| GroupNumber:1 | |
|---|---|
| mt:1 binaryRep: | TickedBool: |
| mt:3 binaryRep: | TickedBool: |
| GroupNumber:1 | |
| mt:2 binaryRep: | TickedBool: |
| mt:3 binaryRep: | TickedBool: |
| GroupNumber:1 | |
| mt:2 binaryRep: | TickedBool: |
| mt:6 binaryRep: | TickedBool: |
| GroupNumber:2 | |
| mt:3 binaryRep: | TickedBool: |
| mt:7 binaryRep: | TickedBool: |
| GroupNumber:2 | |
| mt:3 binaryRep: | TickedBool: |
| mt:11 binaryRep: | TickedBool: |
| GroupNumber:2 | |
| mt:6 binaryRep: | TickedBool: |
| mt:7 binaryRep: | TickedBool: |

Table .1.16.3

In Table .1.16.2, for every value that is used in Table .1.16.3, a tick mark is put beside the midterm value, giving Table .1.16.4.

| GroupNumber:1 | | |
|---|---|---|
| mt:1 binaryRep:1000 | TickedBool: | ✓ |
| GroupNumber:1 | | |
| mt:2 binaryRep:0100 | TickedBool: | ✓ |
| GroupNumber:2 | | |
| mt:3 binaryRep:1100 | TickedBool: | ✓ |
| GroupNumber:2 | | |
| mt:6 binaryRep:0110 | TickedBool: | ✓ |
| GroupNumber:3 | | |
| mt:7 binaryRep:1110 | TickedBool: | ✓ |
| GroupNumber:3 | | |
| mt:11 binaryRep:1101 | TickedBool: | ✓ |

Table .1.16.4

From Table.1.16.3, the nth midterm group is compared with the nth + 1 group. This gives Table.1.16.5, where the first and second midterm differ by a power of 2 and where the second midterm and the last differ by a power of 2.

| GroupNumber:1 | |
|---|---|
| mt:2 binaryRep: | TickedBool: |
| mt:3 binaryRep: | TickedBool: |
| mt:6 binaryRep: | TickedBool: |
| mt:7 binaryRep: | TickedBool: |
| GroupNumber:1 | |
| mt:2 binaryRep: | TickedBool: |
| mt:6 binaryRep: | TickedBool: |
| mt:3 binaryRep: | TickedBool: |
| mt:7 binaryRep: | TickedBool: |

Table .1.16.5

Table.1.16.3 is now ticked for midterm values used in Table.16.5, giving it again (Table .1.16.6) with ticked values

| GroupNumber:1 | | |
|---|---|---|
| mt:1 binaryRep: | TickedBool: | |
| mt:3 binaryRep: | TickedBool: | |
| GroupNumber:1 | | |
| mt:2 binaryRep: | TickedBool: | ✓ |
| mt:3 binaryRep: | TickedBool: | ✓ |
| GroupNumber:1 | | |
| mt:2 binaryRep: | TickedBool: | ✓ |
| mt:6 binaryRep: | TickedBool: | ✓ |
| GroupNumber:2 | | |
| mt:3 binaryRep: | TickedBool: | ✓ |
| mt:7 binaryRep: | TickedBool: | ✓ |
| GroupNumber:2 | | |
| mt:3 binaryRep: | TickedBool: | |
| mt:11 binaryRep: | TickedBool: | |
| GroupNumber:2 | | |
| mt:6 binaryRep: | TickedBool: | ✓ |
| mt:7 binaryRep: | TickedBool: | ✓ |

Table .1.16.6

From Tables Table .1.16.6 and Table .1.16.4 and Table 1.16.5, the unticked midterms are put into the prime implicants term table, where the value in the brackets beside the midterms (in the decimal column)the 1 in the binary representation of the value in the brackets beside the decimal lists is ignored in the binary column to generate the term.For example decimal 1,3(2) = 1100.

Decimal 2 is 01 in binary so in the binary value the Term is read 1_00 where B2 is ignored and the Term is B1A1'A2'.(reading LSB to MSB).

| Prime Implicants Term Table | | |
|---|---|---|
| Decimal | Binary(B1B2A1A2) | Term |
| 1,3,(2) | 1100 | B1A1'A2' |
| 3,11,(8) | 1101 | B1B2A1' |
| 2,3,6,7,(1,4) | 1110 | B2A2' |
| 2,6,3,7,(4,1) | 1110 | B2A2' |

Table .1.16.7

Notice in the prime implicants term table that there is a duplicate term. This can be ignored. Thus, the corrected table is T.1.16.8

| Prime Implicants Term Table | | |
|---|---|---|
| Decimal | Binary(B1B2A1A2) | Term |
| 1,3,(2) | 1100 | B1A1'A2' |
| 3,11,(8) | 1101 | B1B2A1' |
| 2,3,6,7,(1,4) | 1110 | B2A2' |

Table .1.16.8

From Table.1.16.8 and the decimal values of the truth table for the selected function A < B, the prime implicants table can be generated. In this table (Table.1.16.9), the rows with a single X are ticked, and then the midterm values in these rows the columns of the midterms are ticked. Then a best fit for the row to unticked columns is found, and the row and columns are ticked. For a complete solution, all the columns need to be ticked.

| Prime Implicants Table | | | | | | |
|---|---|---|---|---|---|---|
| | 3 | 11 | 7 | 6 | 2 | 1 |
| ✓ B1A1'A2' 1,3,| | X | | | | | X |
| ✓ B1B2A1' 3,11,| | X | X | | | | |
| ✓ B2A2' 2,3,6,7,| | X | | X | X | X | |
| | | | | | | |
| | ✓ | ✓ | ✓ | ✓ | ✓ | ✓ |

Table .1.16.9

From Table.1.16.9, it can be seen that the final resulting expression is below:

Result Expression: B1A1'A2' + B1B2A1' + B2A2'

From this expression, the circuit can be drawn up as shown in Fig. 1.16.1:

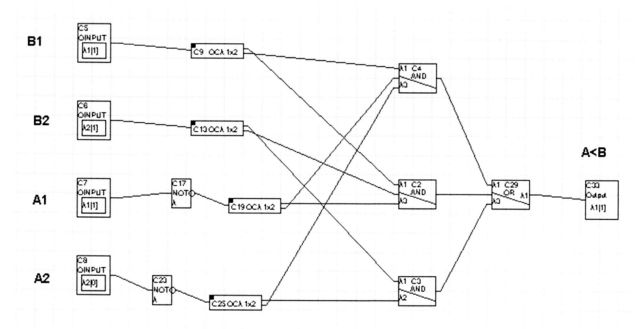

Fig. 1.16.1. Magnitude comparator, A < B circuit

I reran the circuit solver for A = B and got the following result:

Result Expression: B1'B2'A1'A2' + B1B2'A1A2' + B1'B2A1'A2 + B1B2A1A2

This is the following circuit, Fig. 1.16.2:

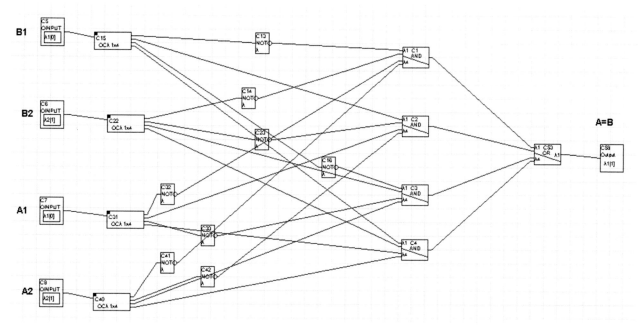

Fig. 1.16.2. Magnitude comparator for A = B

Again, I reran the circuit solver for A > B and got the following result:

Result Expression: B1'B2'A1 + B1'A1A2 + B2'A2

This generates the following circuit, Fig. 1.16.3:

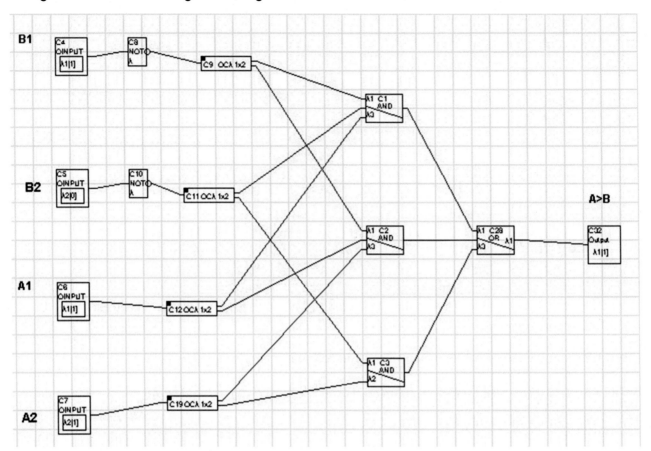

Fig. 1.16.3. Magnitude comparator for A > B

With the three functions A < B, A = B, and A > B, it is possible to generate a circuit that can be further developed to include A <= B and A >= B as shown in the block diagram below.

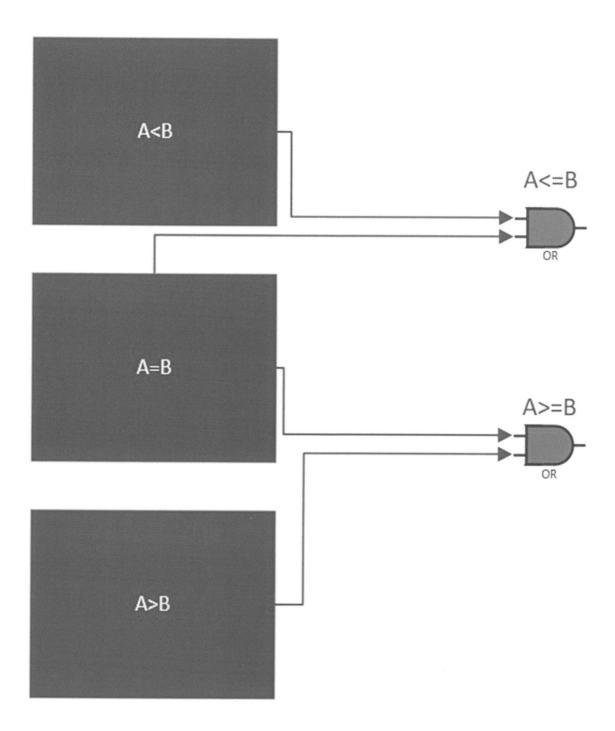

Fig. 1.16.4. A four-block model magnitude comparator

# 1.17. ALU Core Theory

An arithmetic logic unit is just a multiplexed function block with a register for input A and input B, and the result where, with the Fselect, a function 1 to N can be chosen to operate on a data set (Fig. 1.17.0).

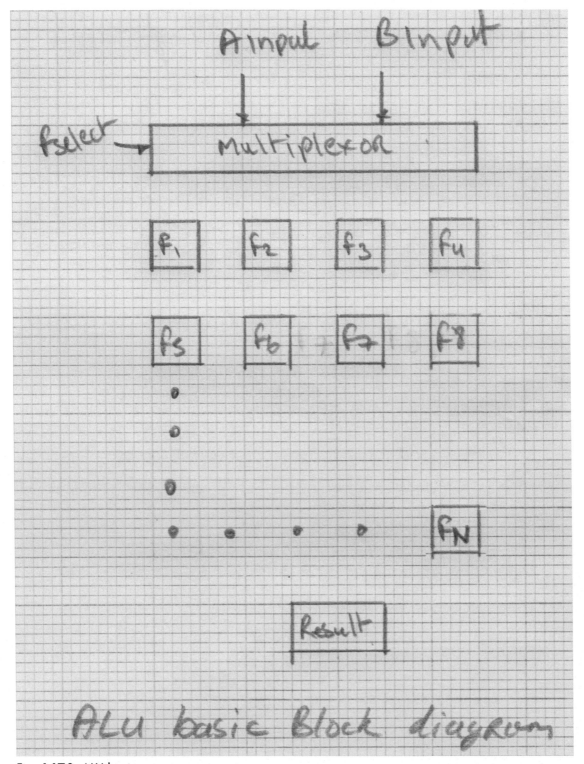

Fig. 1.17.0. ALU basics

# Introduction to Parallelism within a Core

I have shown several functions for an optical processor. Now I am going to introduce the high-level architecture of an optical core. First I will be introducing attempts at parallelism, which are pipelining and threading.

I am first going to introduce pipelining.

Pipelining is a method used to introduce a multi-agent assembly line into a core in order to give a magnitude of parallelism.

Let's divide the processor into sections which are latched for a five-stage pipeline. The stages are as follows:

1.  Fetch instruction from memory and update the PC (IF).
2.  Decode the instruction (ID).
3.  Execute instruction (EX).
4.  Memory access. Load/Store value or else do nothing (MEM).
5.  Write back. Write result into register file if instruction not a branch or store operation (WB).

Fig. 1.17.1 shows an abstract model of a core with pipelining.

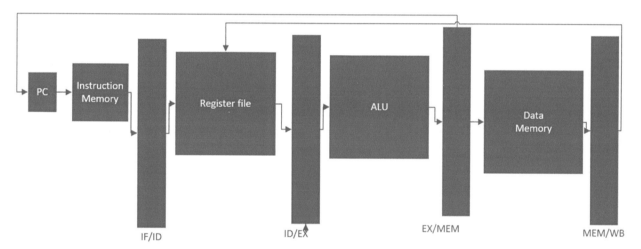

Fig. 1.17.1. Basic pipelining

With pipelining, there are problems. Issues being

1. data hazard RAW (read after write),
2. data hazard WAW (write after write),
3. data hazard WAR (write after read),
4. control hazard/branching,
5. pipeline stall.

Items 1,2, and 3 can be dealt with via forwarding. Data forwarding is a method used in pipelined computers to limit the amount of pipeline stalls.

Control hazard/branching are delays in determining the proper instruction to fetch. To reduce the time penalty for this type of hazard, it is possible to move the conditional branch execution to the ID stage. This will mean that on hazard, only one time delay is experienced—that being of the current branch instruction—as a penalty.

A pipeline stall is where a hazard occurs which causes a delay in order to resolve the hazard.

# 1.17.1. Threading

I am now going to introduce the threading model for the various cores. But before that, I am going to introduce some concepts.

First, when a scene is loaded, each thread is given a kernel in order for it to be able to do a particular simulation. The threads are used in a streaming fashion—where a thread gets data, computes values from it, and then puts the results back into memory. That is how the threads are used on the main simulation node, which is a data-centre-scale simulation where the scene test case is 100 km × 100 km × 100 km. But before I deal with the scene, the threads on the holodeck display unit get data compute values and store the results into the settings memory for the drivers for the emitters, the settings being solidness; x, y, z moment angles; and magnitude of the moment.

I will explain. Solidness is the amount in newtons of resistance force to touch at that point. X, y, z moment angles are used to set up a moment on a sphere. The magnitude of the moments could be seen to be the solidness felt at the face of the triangulated moments. These settings would vary for particular materials with various colours and textures.

At the end of the volumetrics section, I introduce several attempts at creating solidness for a virtual material and how, if the electromagnetic fields don't achieve the effect, then it is also possible to set up other field patterns which would interact with the parts of an atom in order to create solidness.

I am now going to jump ahead and introduce a personal physical volume (Ppv), which is a kind of personal holodeck which is 4 meters × 4 meters × 4 meters in volume. Within this volume, I have the resolution of the voxels to be 20 micrometers ($20 \times 10^{-6}$ metres). I need to introduce some math now in order to explain further.

There are $1 / 20 \times 10^{-6} = 50,000$ voxels of width 20 μm in 1 meter.

50,000 × 4 = 200,000 voxels in 4 meters.

200,000 × 200,000 × 200,000 = 8 × 10$^{15}$ voxels within the Ppv volume.

Each of these voxels has 1000 force points—10 layers of 10 × 10 = 1000.

A thread takes care of ten of these force points where the thread has to compute lighting, sound, and solidness; and the core has to rasterise the voxel with the force points, giving a face to triangular data. Where a line could be envisioned by creating a material of a certain type with solidness and with three force points as a triangle face, a point for the line could be seen where the magnetic moments would be set up to reflect a ray of light at the colour of the material of the line could be seen if one could see at this level, but Fig. 1.17.1.1 should clarify the details.

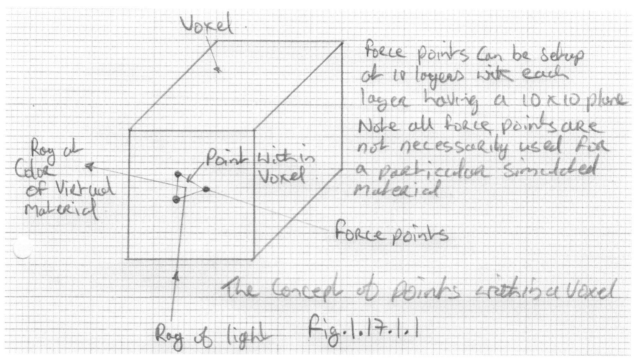

Fig. 1.17.1.1. The concept of points within a voxel

The threads that deal with the force points in Fig. 1.17.1.1 have to calculate the magnetic moment and pass the angle information in three dimensions to the emitter, along with the magnitude of the moment needed for each point. By varying the width between the force points and the magnitude of the moments, different colours of the ray could be envisioned.

A core/voxel is made up of 101 threads where each thread except the first (main thread which takes care of storing of buffer info, etc.) takes care of 10 force points. A force point is a point within a voxel where an electromagnetic field can be set up in order for solidness to be experienced or where the force point is part of a field pattern in order to bend rays of light to give the effect of reflection. (Note that the direction vector of the ray has to be changed to the new reflected direction. More on this in the holodeck section of the book.) Fig. 1.17.1.2 shows a high-level diagram of a core where each core deals with a voxel so the mapping is fairly straightforward. In the diagram, I show that each thread has a branch unit, floating point unit, integer unit, and a load/store unit, and each thread is pipelined. As explained above, the main thread of every core is used for various functions, but a very important function is that each thread has a link to the buffer. The buffer at a core level will be for thread-level information to make up a core packet to be sent or received from the main subcell threading unit, where the buffer there holds the information for a packet of information to hold the information for the subcell grid unit, as shown in the XML below. (To optimise this XML, it could be better to arrange the data into a bit-level packet; but for clarity, XML is used).

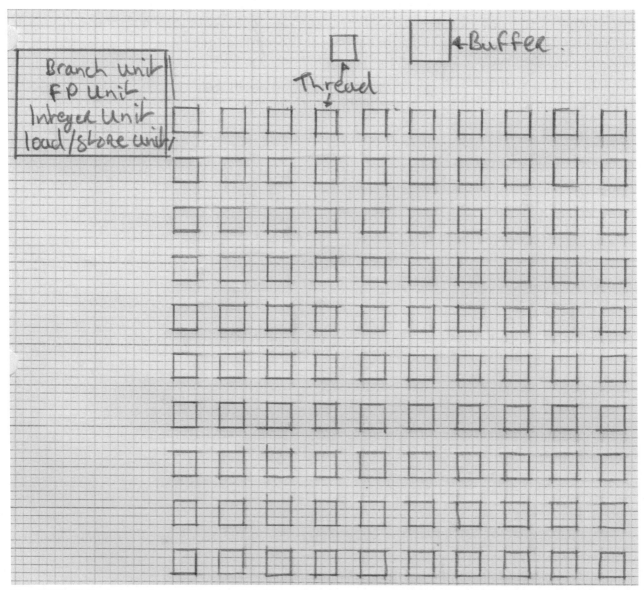

Fig. 1.17.1.2. Basic core layout showing threads and the functions within the threads (branch unit, floating point unit, integer unit and load/store unit)

From Fig. 1.17.1.2

- A branch unit is just a block of circuitry that deals with branches.
- A floating point unit (64-bit) is a block that deals with floating point instructions.
- An integer unit (32-bit) is a block that deals with integer instructions.
- A load/store unit (128-bit) deals with loading and storing of information.

Before I show the XML, I am going to introduce the concept of a subcell grid. In order to realistically load a scene at 60 Hz, the scene is transferred from the data centre to the Ppv via fibre-optic cable. There is a certain threshold of information that can be processed at a fraction of a second, so since the fastest connection speed in 2014 was about 34 TBs per second, I chose to have a volume of $1 \times 10^{12}$ voxels to represent the information that needs to be sent down the fibre-optical cable on a fraction-of-a-second level. I have to do some math here.

With $8 \times 10^{15} / (10 \times 1000)^3 = 8000$, a personal physical volume broken down into 1 teravoxels needs 8000 optical cables to move data from RAM in the data centre to/from RAM in the volumetric display unit Ppv at about 60 Hz. That means that each strand is capable of terabytes per fraction of a second.

For the optical core configuration, the buffer on the core will hold the thread-level information shown here in XML, where each thread has a fiber link to its own memory locations within the buffer so each thread can save its data concurrently.

<core number>

    <thread number>

        Force point number; solidness; x, y, z moment angles; moment magnitude

        Force point number; solidness; x, y, z moment angles; moment magnitude

        ....

    </thread number>

    ....

</core number>

Note this: from above, if a force point is turned off (as it might not be needed to simulate a particular material) then the tristate switch connecting the thread to the buffer is turned off (dark). If the thread/force point is used, then the tristate switch can be either high or low. It is important to remember that 10 force points are dealt with per thread; thus, if only some are used, the thread information is only the used information. Also note that only changes are sent/received.

For the core that deals with the subcell grid number 1 of the 8000 subcell grid numbers, the buffer there is filled with core-level information (XML shown below). What I mean by core level is that each core concurrently adds its core-level information to the buffer to build the subcell grid number packet before it is sent to the Ppv number for rendering.

    <subcell gridnumber>

        <core number>

        <thread number>

            Thread level info as before

        </thread number>

        ...

        <Texture number>

N u,v

</Texture number>

<update frequency for force points or spatial light modulator>

</core number>

.....

<core>

...

</core>

..

<texture image>

</subcell gridnumber>

</Ppv number>

I mentioned above that the test case scene on the main simulation node (data centre) is 100 km × 100 km × 100 km. On this scene, each Ppv can move virtually through the scene, incrementing in any given direction a step equivalent to 10,000, the subcell grid number of voxels in any given axis (10,000 × 10,000 × 10,000 = $1 \times 10^{12}$, which is the number of voxels within the subcell grid). Getting back to the $10,000 \times 20 \times 10^{-6} = 0.2$ meters (approximately 7 7/8 inches), this happens to be about a step. So it works out very well for snapping the Ppv to the subcell grid, as it scrolls the data centre scene. This could be adjusted to an integer number of subcell grids (0.2 m) if your step is larger.

The texture image is the Ppv sphere which is on all multi-user simulations in the Ppv. It is automatically generated on the fly. And if the sub-grid in question lies on part of the Ppv sphere, the autogenerated and level-of-detail texture is referenced with coordinates like north, south, east,west,top, bottom, and then the UV coordinates. The directions refer to the part of the voxel to be faced with the texture.

Now that all of that is covered, I am going to cover what happens when you load a simulation. First, I have the scene fragmented into subcell grids so it loads $1 \times 10^{12}$ voxels of information at a time from optical storage discs. Since the scene is so large, the discs are RAID-controlled. Fig. 1.17.1.3 shows the core with the RAID-controlled optical discs. This is similar to the multi-user set-up described above, where for quick loading, I chose 1 teravoxel as a magic number to load into the holodeck's RAM subcell grid at a time. Each subcell grid has a similar set-up, and the full scene is the combined efforts of loading each subcell grid concurrently. Thus, the full scene should load in the matter of minutes (data centre scene 100 km × 100 km × 100 km).

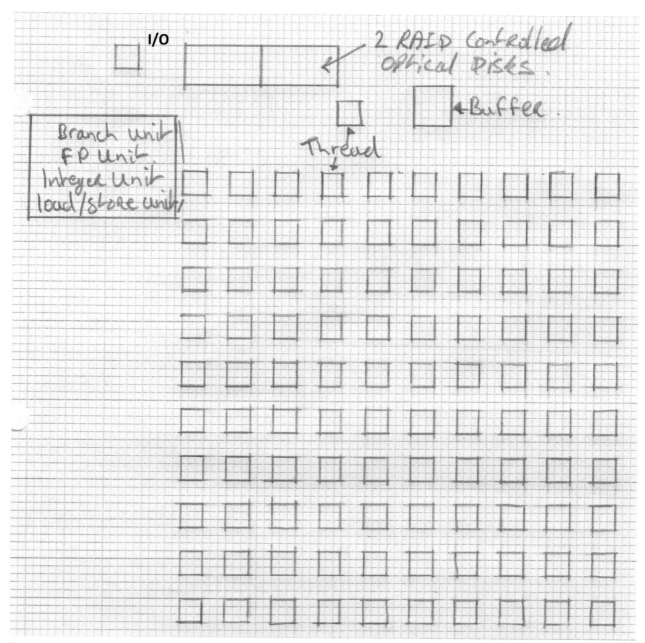

Fig. 1.17.1.3. Core with RAID-controlled optical discs.

From Fig. 1.17.1.3

- A branch unit is just a block of circuitry that deals with branches.
- A floating point unit (64-bit) is a block that deals with floating point instructions.
- An integer unit (32-bit) is a block that deals with integer instructions.
- A load/store unit (128-bit) deals with loading and storing of information.

The volumetric display unit for home users (which works through a networked environment) can be used as a standalone static holodeck with the use of its built-in storage or the special file server that will be given with the holodeck unit. The interface for the file server will be similar to the networked set-up where an 8000-strand optical fibre trunk will be connected to the holodeck input/output port. It is important to note here that each core for the subcell grid has an input/

output port, and one strand of the trunk is connected to the subcell grid unit, but there are 8000 subcell grid units for a Ppv.

Each thread in the unit has its own memory for programming/kernel, and it shares certain data with its nearest neighbours in 3D. So for a normal cell not in the corner or at the edge, it will have nine top nearest neighbours; in the middle, eight nearest neighbours; and at the bottom, nine nearest neighbours. For the Ppv thread, level information is shared for streaming on moving. For the data centre, each thread has access to it own share of global memory, and it may need a similar sharing approach for streaming movements.

I am now going to discuss buffering within the volumetric display unit. In the VDU, there could be a buffer on a per-thread basis for each force point for information like

1. solidness (force in newtons),
2. x, y, z moment angles,
3. moment magnitude,
4. set on a per-voxel basis the frequency of update.

The emitters take information from these buffers to adjust the VDU, so to stop flicker, a buffer with these values in it could be used and swapped when needed (when calculation is finished).

Loading a scene is quite complex in a data centre, as the scene is 100 km x 100 km x 100 km in virtual space at a resolution of 20 μm. Every grid unit has a board for loading/storing a grid unit's worth of a scene (Fig. 1.17.1.3). Thus the whole data centre scene is fragmented into parts which are concurrently loaded by each subcell grid unit from optical storage. So the scene as a whole should load within minutes. A similar procedure is done for saving a scene. Every subcell grid unit has a core with optical discs for storage, and the subcell grid unit saves the scene from the concurrently created buffer data to the optical disc.

# Holodeck Motherboard Theory

In this section I roughly sketch out how the motherboards will come together by layering. I also show the fundamentals of how the concept of multi-user holodecks function.

## 1.18.1. Data Centre Holodeck Motherboard Rough Requirements

For the data centre, a massively parallel motherboard is required. The scene is 100 km x 100 km x 100 km in virtual space. So let's do the math.

$1/20 \times 10^{-6} = 50,000 = 50K$ voxels per metre

$100K \times 50K = 5 \times 10^9$ voxels per 100 km

$5 \times 10^9 \times 5 \times 10^9 \times 5 \times 10^9 = 1.25 \times 10^{29}$ voxels per volume

$1.25 \times 10^{29} / ((10 \times 1000)^3) = 1.25 \times 10^{17}$ optical links or subcell grid units (Fig. 1.18.1.1)

= 500 × 10³ grid units per axis on the motherboard (Fig. 1.18.1.3), where the grid units are shown in Fig. 1.18.1.2.

A subcell grid unit is made up of 10,000 × 10,000 × 10,000 voxels where, on the layered board, each core maps to a voxel of information.

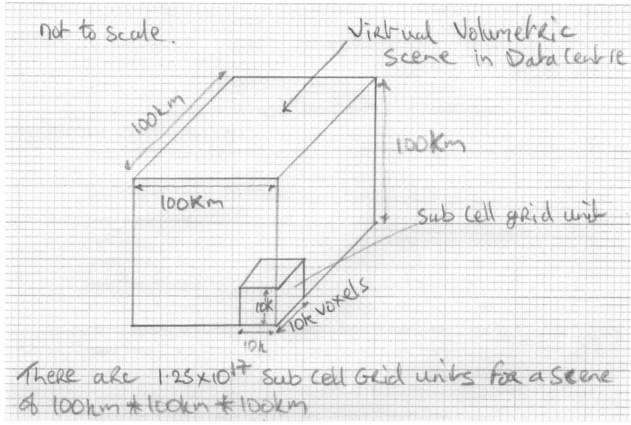

Fig. 1.18.1.1. Virtual data centre scene showing a subcell grid unit in voxels

Every core has 100 threads to simulate the scene.

That's 1.25 × 10²⁹ × 100 = 1.25 × 10³¹ threads, and each thread for this test case takes care of ten force points. Also, each thread has its own share of global memory, say 500 KB.

We would require 1.25 × 10³¹ × 500 × 1024 = 6.4 × 10³⁶ bytes of memory for the data centre simulation engine. This gives a thread kernel a lot of space of calculation where it needs to be able to calculate light (reflection model); sound; level of detailed texture; moment angles x, y, z of the force points; magnitude of the moments; and solidness of the simulated material. Fig. 1.18.1.2 shows the actual mappings of cores within a subcell grid unit where the unit is layered in motherboard space.

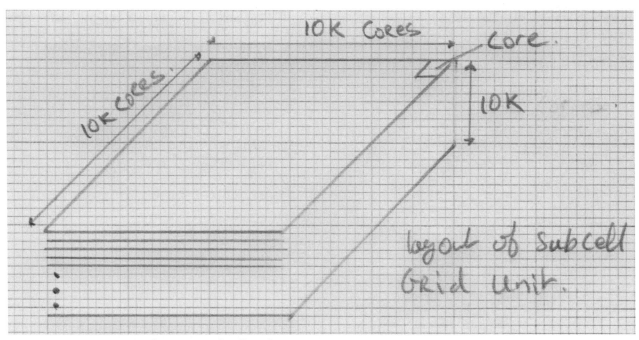

Fig. 1.18.1.2. Mappings of cores in subcell grid unit

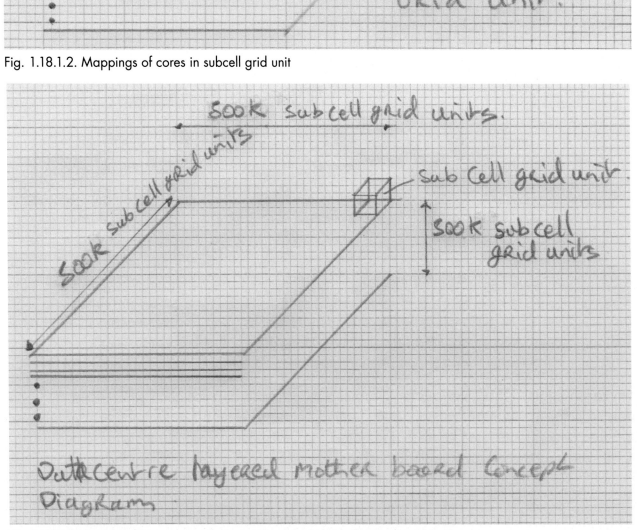

Fig. 1.18.1.3. Data centre concept of a layered motherboard of subcell grid units

It should be noted that the amount of layers will be determined by cooling needs. So gaps may be needed within the layers.

For the holodeck to function correctly, the data centre is gridded with subcell grid units. When a person moves within, say, Ppv1 (Fig. 1.18.1.4), the scene they see is created in virtual space within the data centre's virtual global scene. When the person moves within the Ppv1 scene, the data centre's scene moves almost around them, scrolling in increments of a side of a grid unit in virtual space (integer increments of 0.2 m, as explained before). The router directs the volume taken up by virtual Ppv1 to actual, real Ppv1. This routing has to route each of the subcell grid units taken up by the VPpv1 (8000) to the real Ppv1.

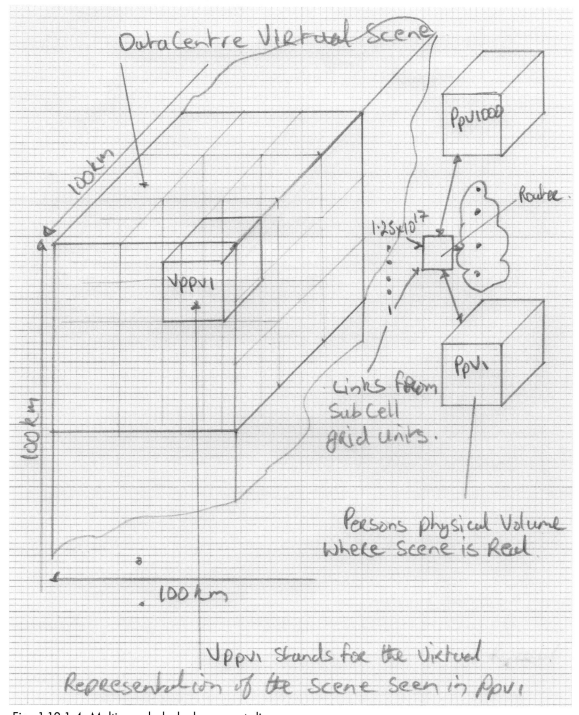

Fig. 1.18.1.4. Multi-user holodeck concept diagram

## 1.18.2. Personal Holodeck Ppv Motherboard Rough Requirements

Personal physical volumes (Ppv) are now covered. I am again going to break the volume of the volumetric display into subcell grid units and then show the mappings of the grid units on a layered motherboard.

First, let's do the math.

$1 / 20 \times 10^{-6}$ = 50,000 or 50K voxels in 1 metre.

$4 \times 50K = 200 \times 10^3$ voxels in 4 metres.

$200K \times 200K \times 200K = 8 \times 10^{15}$ voxels per Ppv.

$8 \times 10^{15} / ((10 \times 1000)^3)$ = 8000 optical links or subcell grid units.

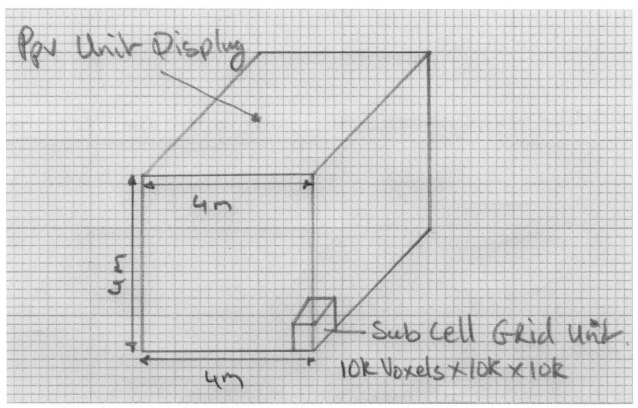

Fig. 1.18.2.1. Ppv display showing concept of subcell grid unit

$\sqrt[3]{8000} = 20$

This means there are twenty grid units per axis.

Fig. 1.18.2.2 shows a concept motherboard where Fig. 1.18.1.2 shows the core mappings within a subcell grid unit.

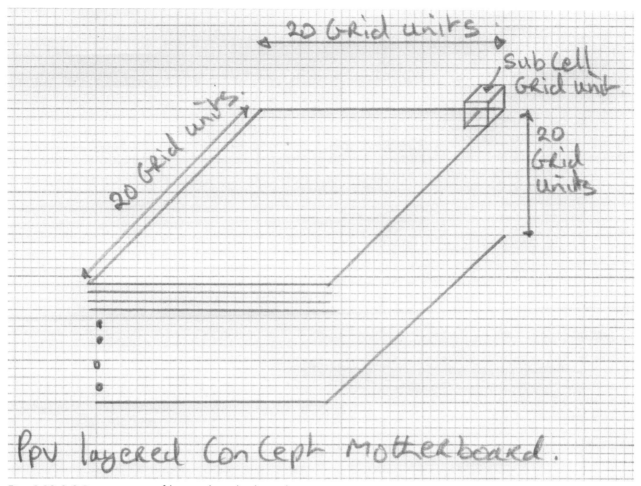

Fig. 1.18.2.2 Ppv concept of layered motherboard

Some of the thread memory is shared for variables like solidness; x, y, z moment angles; moment magnitude; sound level; etc. These variables are stored in set-aside addresses, where the settings are used in the driver to drive the emitters. Also there are two set-aside locations for these variables—one for the first VDU settings online buffer and the other for the offline buffer. These can be easily swapped. Thus, this is the double-buffer mechanism.

It is worth noting the amount of rendering threads in the VDU (volumetric display unit).

$8 \times 10^{15} \times 100 = 8 \times 10^{17}$ threads in Ppv unit.

Each thread manages ten force points.

$8 \times 10^{17} \times 10 = 8 \times 10^{18}$ total force points.

Each thread has access to its own memory, where some of the addresses are used for shared storage as mentioned above. Let's say for the data centre threads that each thread has 500KB of memory.

$8 \times 10^{17} \times 500 \times 1024 = 4.096 \times 10^{23}$ bytes of memory used in a Ppv.

Loading a static scene is quite complex in a Ppv, as the scene is 4 m $\times$ 4 m $\times$ 4 m at a resolution of 20 μm. Every grid unit has a board for loading/storing a grid unit's worth of a scene (Fig. 1.17.1.3). Thus the whole scene is fragmented into parts which are loaded by each subcell grid unit from optical storage concurrently (8000 of them). So the scene as a whole should load within minutes. A similar procedure is done for saving a scene. Every subcell grid unit has a core with optical discs for storage, and the subcell grid unit saves the scene from the concurrently created buffer data to the optical discs.

# Part 2

2.1.0. Instruction Set Theory Introduction

2.1.1. Instruction Set Listing for RISC-V

2.1.2. Theory and Notes

## 2.1.0. Instruction Set Theory

In this part of the book, I am going to introduce the RISC-V open instruction set architecture. I chose RISC-V as its open, free, modifiable, and has what I need in it. With this instruction set, it is possible to use the quad precision standard and thus address $2^{128}$ addresses. This is approximately $3.4 \times 10^{38}$ addresses. I also map part 1 hardware to possible instructions via modifying the hardware.

## 2.1.1. Instruction Set Listing for RISC-V

| ʌ32 ʌ26 | ʌ25 ʌ21 | ʌ20 ʌ16 | ʌ15 ʌ13 | ʌ12 ʌ8 | ʌ7 ʌ1 | |
|---|---|---|---|---|---|---|
| Funct7 | Rs2 | Rs1 | Funct3 | rd | opcode | R-type |
| Imm[11:0] | | Rs1 | Funct3 | rd | opcode | I-type |
| Imm[11:5] | Rs2 | Rs1 | Funct3 | Imm[4:0] | opcode | S-type |
| Imm[12|10:5] | Rs2 | Rs1 | Funct3 | Imm[4:1|11] | opcode | B-type |
| Imm[31:12] | | | | rd | opcode | U-type |
| Imm[20|10:1|11|19:12] | | | | rd | opcode | J-type |

Table 2.1.1.1. Instruction type frames

## RV32I Base Instruction Set

| | | | | | | |
|---|---|---|---|---|---|---|
| Imm[31:12] | | | | rd | 0110111 | LUI |
| Imm[31:12] | | | | rd | 0010111 | AUIPC |
| Imm[20|10:1|11|19:12] | | | | Rd | 1101111 | JAL |
| Imm[11:0] | | Rs1 | 000 | Rd | 1100111 | JALR |
| Imm[12|10:5] | Rs2 | Rs1 | 000 | Imm[4:1|11] | 1100011 | BEQ |
| Imm[12|10:5] | Rs2 | Rs1 | 001 | Imm[4:1|11] | 1100011 | BNE |
| Imm[12|10:5] | Rs2 | Rs1 | 100 | Imm[4:1|11] | 1100011 | BLT |
| Imm[12|10:5] | Rs2 | Rs1 | 101 | Imm[4:1|11] | 1100011 | BGE |
| Imm[12|10:5] | Rs2 | Rs1 | 110 | Imm[4:1|11] | 1100011 | BLTU |
| Imm[12|10:5] | Rs2 | Rs1 | 111 | Imm[4:1|11] | 1100011 | BGEU |
| Imm[11:0] | | Rs1 | 0000 | Rd | 0000011 | LB |
| Imm[11:0] | | Rs1 | 001 | Rd | 0000011 | LH |
| Imm[11:0] | | Rs1 | 010 | Rd | 0000011 | LW |
| Imm[11:0] | | Rs1 | 100 | Rd | 0000011 | LBU |
| Imm[11:0] | | Rs1 | 101 | Rd | 0000011 | LHU |
| Imm[11:5] | Rs2 | Rs1 | 000 | Imm[4:0] | 0100011 | SB |
| Imm[11:5] | Rs2 | Rs1 | 001 | Imm[4:0] | 0100011 | SH |
| Imm[11:5] | Rs2 | Rs1 | 010 | Imm[4:0] | 0100011 | SW |
| Imm[11:0] | | Rs1 | 000 | Rd | 0010011 | ADDI |
| Imm[11:0] | | Rs1 | 010 | Rd | 0010011 | SLTI |
| Imm[11:0] | | Rs1 | 011 | Rd | 0010011 | SLTIU |

| | | | | | | |
|---|---|---|---|---|---|---|
| Imm[11:0] | | Rs1 | 100 | Rd | 0010011 | XORI |
| Imm[11:0] | | Rs1 | 110 | Rd | 0010011 | ORI |
| Imm[11:0] | | Rs1 | 111 | Rd | 0010011 | ANDI |
| 0000000 | Shamt | Rs1 | 001 | Rd | 0010011 | SLLI |
| 0100000 | Shamt | Rs1 | 101 | Rd | 0010011 | SRLI |
| 0000000 | shamt | Rs1 | 101 | Rd | 0010011 | SRAI |
| 0000000 | Rs2 | Rs1 | 000 | Rd | 0110011 | ADD |
| 0100000 | Rs2 | Rs1 | 000 | Rd | 0110011 | SUB |
| 0000000 | Rs2 | Rs1 | 001 | Rd | 0110011 | SLL |
| 0000000 | Rs2 | Rs1 | 010 | Rd | 0110011 | SLT |
| 0000000 | Rs2 | Rs1 | 011 | Rd | 0110011 | SLTU |
| 0000000 | Rs2 | Rs1 | 100 | Rd | 0110011 | XOR |
| 0000000 | Rs2 | Rs1 | 101 | Rd | 0110011 | SRL |
| 0100000 | Rs2 | Rs1 | 101 | Rd | 0110011 | SRA |
| 0000000 | Rs2 | Rs1 | 110 | Rd | 0110011 | OR |
| 0000000 | Rs2 | Rs1 | 111 | Rd | 0110011 | AND |
| fm | pred | Succ | Rs1 | 000 | Rd | 0001111 | FENCE |
| 000000000000 | | 00000 | 000 | 00000 | 1110011 | ECALL |
| 000000000001 | | 00000 | 000 | 00000 | 1110011 | EBREAK |

Table 2.1.1.2. RV32I instruction frames

| λ32 λ26 | λ25 λ21 | λ20 λ16 | λ15 λ13 | λ12 λ8 | λ7 λ1 | |
|---|---|---|---|---|---|---|
| Funct7 | Rs2 | Rs1 | Funct3 | rd | opcode | R-type |
| Imm[11:0] | | Rs1 | Funct3 | rd | Opcode | I-type |
| Imm[11:5] | Rs2 | Rs1 | Funct3 | Imm[4:0] | opcode | S-type |

Table 2.1.1.3. RV64I RV32/RV64 Zifencei RV21/RV64 Zicsr RV32M RV64M instruction type frames

## RV64I Base Instruction Set (in addition to RV32I)

| Imm[11:0] | | Rs1 | 110 | Rd | 0000011 | LWU |
|---|---|---|---|---|---|---|
| Imm[11:0] | | Rs1 | 011 | Rd | 0000011 | LD |
| Imm[11:5] | Rs2 | Rs1 | 011 | Imm[4:0] | 0100011 | SD |
| 000000 | Shamt | Rs1 | 001 | Rd | 0010011 | SLLI |
| 000000 | Shamt | Rs1 | 101 | Rd | 0010011 | SRLI |
| 010000 | shamt | Rs1 | 101 | Rd | 0010011 | SRAI |
| Imm[11:0] | | Rs1 | 000 | Rd | 0011011 | ADDIW |
| 0000000 | Shamt | Rs1 | 001 | Rd | 0011011 | SLLIW |
| 0000000 | Shamt | Rs1 | 101 | Rd | 0011011 | SRLIW |
| 0100000 | Shamt | Rs1 | 101 | Rd | 0011011 | SRAIW |
| 0000000 | Rs2 | Rs1 | 000 | Rd | 0111011 | ADDW |
| 0100000 | Rs2 | Rs1 | 000 | Rd | 0111011 | SUBW |
| 0000000 | Rs2 | Rs1 | 001 | Rd | 0111011 | SLLW |
| 0000000 | Rs2 | Rs1 | 101 | Rd | 0111011 | SRLW |
| 0100000 | Rs2 | Rs1 | 101 | Rd | 0111011 | SRAW |

Table 2.1.1.4. RV64i instruction frames

## RV32/RV64 Zifencei Standard Extension

| Imm[11:0] | Rs1 | 001 | rd | 0001111 | FENCE.I |
|---|---|---|---|---|---|

Table 2.1.1.5. RV32/RV64 Zifencei instruction frame

## RV32/RV64 Zicsr Standard Extension

| Csr | Rs1 | 001 | Rd | 1110011 | CSRRW |
|---|---|---|---|---|---|
| Csr | Rs1 | 010 | Rd | 1110011 | CSRRS |
| Csr | Rs1 | 011 | Rd | 1110011 | CSRRC |
| Csr | Uimm | 101 | Rd | 1110011 | CSRRWI |
| Csr | Uimm | 110 | Rd | 1110011 | CSRRSI |
| Csr | Uimm | 111 | Rd | 1110011 | CSRRCI |

Table 2.1.1.6. RV32/64 Zicsr instruction frames

## RV32M Standard Extension

| 0000001 | Rs2 | Rs1 | 000 | Rd | 0110011 | MUL |
|---|---|---|---|---|---|---|
| 0000001 | Rs2 | Rs1 | 001 | Rd | 0110011 | MULH |
| 0000001 | Rs2 | Rs1 | 010 | Rd | 0110011 | MULHSU |
| 0000001 | Rs2 | Rs1 | 011 | Rd | 0110011 | MULHU |
| 0000001 | Rs2 | Rs1 | 100 | Rd | 0110011 | DIV |
| 0000001 | Rs2 | Rs1 | 101 | Rd | 0110011 | DIVU |
| 0000001 | Rs2 | Rs1 | 110 | Rd | 0110011 | REM |
| 0000001 | Rs2 | Rs1 | 111 | Rd | 0110011 | REMU |

Table 2.1.1.7. RV32M instruction frames

## RV64M Standard Extension (in addition to RV32M)

| 0000001 | Rs2 | Rs1 | 000 | Rd | 0111011 | MULW |
|---|---|---|---|---|---|---|
| 0000001 | Rs2 | Rs1 | 100 | Rd | 0111011 | DIVW |
| 0000001 | Rs2 | Rs1 | 101 | Rd | 0111011 | DIVUW |
| 0000001 | Rs2 | Rs1 | 110 | Rd | 0111011 | REMW |
| 0000001 | Rs2 | Rs1 | 111 | Rd | 0111011 | REMUW |

Table 2.1.1.8. RV64M instruction frames

| λ32 | λ28 | λ27λ26 | λ25λ21 | λ20λ16 | λ15 13 | λ12λ8 | λ7 λ1 | |
|---|---|---|---|---|---|---|---|---|
| Funct7 | | | Rs2 | Rs1 | Funct3 | Rd | opcode | R-type |

Table 2.1.1.9. RV32A RV64A instruction type frames

## RV32A Standard Extension

| 00010 | Aq | Rl | 00000 | Rs1 | 010 | Rd | 0101111 | LR.W |
|---|---|---|---|---|---|---|---|---|
| 00011 | Aq | Rl | Rs2 | Rs1 | 010 | Rd | 0101111 | SC.W |
| 00001 | Aq | Rl | Rs2 | Rs1 | 010 | Rd | 0101111 | AMOSWAP.W |
| 00000 | Aq | Rl | Rs2 | Rs1 | 010 | Rd | 0101111 | AMOADD.W |
| 00100 | Aq | Rl | Rs2 | Rs1 | 010 | Rd | 0101111 | AMOXOR.W |
| 01100 | Aq | Rl | Rs2 | Rs1 | 010 | Rd | 0101111 | AMOAND.W |
| 01000 | Aq | Rl | Rs2 | Rs1 | 010 | Rd | 0101111 | AMOOR.W |
| 10000 | Aq | Rl | Rs2 | Rs1 | 010 | Rd | 0101111 | AMOMIN.W |
| 10100 | Aq | Rl | Rs2 | Rs1 | 010 | Rd | 0101111 | AMOMAX.W |
| 11000 | Aq | Rl | Rs2 | Rs1 | 010 | Rd | 0101111 | AMOMINU.W |
| 11100 | Aq | Rl | Rs2 | Rs1 | 010 | Rd | 0101111 | AMOMAXU.W |

Table 2.1.1.10. RV32A instruction frames

## RV64A Standard Extension (in addition to RV32A)

| | | | | | | | | |
|---|---|---|---|---|---|---|---|---|
| 00010 | Aq | Rl | 00000 | Rs1 | 011 | Rd | 0101111 | LR.D |
| 00011 | Aq | Rl | Rs2 | Rs1 | 011 | Rd | 0101111 | SC.D |
| 00001 | Aq | Rl | Rs2 | Rs1 | 011 | Rd | 0101111 | AMOSWAP.D |
| 00000 | Aq | Rl | Rs2 | Rs1 | 011 | Rd | 0101111 | AMOADD.D |
| 00100 | Aq | Rl | Rs2 | Rs1 | 011 | Rd | 0101111 | AMOXOR.D |
| 01100 | Aq | Rl | Rs2 | Rs1 | 011 | Rd | 0101111 | AMOAND.D |
| 01000 | Aq | Rl | Rs2 | Rs1 | 011 | Rd | 0101111 | AMOOR.D |
| 10000 | Aq | Rl | Rs2 | Rs1 | 011 | Rd | 0101111 | AMOMIN.D |
| 10100 | Aq | Rl | Rs2 | Rs1 | 011 | Rd | 0101111 | AMOMAX.D |
| 11000 | Aq | Rl | Rs2 | Rs1 | 011 | Rd | 0101111 | AMOMINU.D |
| 11100 | Aq | Rl | Rs2 | Rs1 | 011 | Rd | 0101111 | AMOMAXU.D |

Table 2.1.1.11. RV64A instruction frames

| λ32 λ28 | λ27 λ26 | λ25 λ21 | λ20 λ16 | λ15 λ13 | λ12 λ8 | λ7 λ1 | |
|---|---|---|---|---|---|---|---|
| Funct7 | | Rs2 | Rs1 | Funct3 | Rd | Opcode | R-type |
| Rs3 | Funct2 | Rs2 | Rs1 | Funct3 | Rd | Opcode | R4-type |
| Imm[11:0] | | | Rs1 | Funct3 | Rd | Opcode | I-type |
| Imm[11:5] | | Rs2 | Rs1 | Funct3 | Imm[4:0] | Opcode | S-type |

Table 2.1.1.12. RV32F RV64F instruction type frames

## RV32F Standard Extension

| | | | | | | | |
|---|---|---|---|---|---|---|---|
| Imm[11:0] | | | Rs1 | 010 | Rd | 0000111 | FLW |
| Imm[11:5] | | Rs2 | Rs1 | 010 | Imm[4:0] | 0100111 | FSW |
| Rs3 | 00 | Rs2 | Rs1 | Rm | Rd | 1000011 | FMADD.S |
| Rs3 | 00 | Rs2 | Rs1 | Rm | Rd | 1000111 | FMSUB.S |
| Rs3 | 00 | Rs2 | Rs1 | Rm | Rd | 1001011 | FNMSUB.S |
| Rs3 | 00 | Rs2 | Rs1 | Rm | Rd | 1001111 | FNMADD.S |
| 0000000 | | Rs2 | Rs1 | Rm | Rd | 1010011 | FADD.S |
| 0000100 | | Rs2 | Rs1 | Rm | Rd | 1010011 | FSUB.S |
| 0001000 | | Rs2 | Rs1 | Rm | Rd | 1010011 | FMUL.S |
| 0001100 | | Rs2 | Rs1 | Rm | Rd | 1010011 | FDIV.S |
| 0101100 | | 00000 | Rs1 | Rm | Rd | 1010011 | FSQRT.S |
| 0010000 | | Rs2 | Rs1 | 000 | Rd | 1010011 | FSGNJ.S |
| 0010000 | | Rs2 | Rs1 | 001 | Rd | 1010011 | FSGNJN.S |
| 0010000 | | Rs2 | Rs1 | 010 | Rd | 1010011 | FSGNJX.S |
| 0010100 | | Rs2 | Rs1 | 000 | Rd | 1010011 | FMIN.S |
| 0010100 | | Rs2 | Rs1 | 001 | Rd | 1010011 | FMAX.S |
| 1100000 | | 00000 | Rs1 | Rm | Rd | 1010011 | FCVT.W.S |

| 1100000 | 00001 | Rs1 | Rm | Rd | 1010011 | FCVT.WU.S |
|---|---|---|---|---|---|---|
| 1110000 | 00000 | Rs1 | 000 | Rd | 1010011 | FMV.X.W |
| 1010000 | Rs2 | Rs1 | 010 | Rd | 1010011 | FEQ.S |
| 1010000 | Rs2 | Rs1 | 001 | Rd | 1010011 | FLT.S |
| 1010000 | Rs2 | Rs1 | 000 | Rd | 1010011 | FLE.S |
| 1110000 | 00000 | Rs1 | 001 | Rd | 1010011 | FCLASS.S |
| 1101000 | 00000 | Rs1 | Rm | Rd | 1010011 | FCVT.S.W |
| 1101000 | 00001 | Rs1 | Rm | Rd | 1010011 | FCVT.S.WU |
| 1111000 | 00000 | Rs1 | 000 | Rd | 1010011 | FMV.W.X |

Table 2.1.1.13. RV32F instruction frames

## RV64F Standard Extension (in addition to RV32F)

| 1100000 | 00010 | Rs1 | Rm | Rd | 1010011 | FCVT.L.S |
|---|---|---|---|---|---|---|
| 1100000 | 00011 | Rs1 | Rm | Rd | 1010011 | FCVT.LU.S |
| 1101000 | 00010 | Rs1 | Rm | Rd | 1010011 | FCVT.S.L |
| 1101000 | 00011 | Rs1 | Rm | rd | 1010011 | FCVT.S.LU |

Table 2.1.1.14. RV64F instruction frames

| λ32 λ28 | λ27 λ26 | λ25 λ21 | λ20 λ16 | λ15 λ13 | λ12 λ8 | λ7 λ1 | |
|---|---|---|---|---|---|---|---|
| Funct7 | | Rs2 | Rs1 | Funct3 | Rd | Opcode | R-type |
| Rs3 | Funct2 | Rs2 | Rs1 | Funct3 | Rd | Opcode | R4-type |
| Imm[11:0] | | | Rs1 | Funct3 | Rd | Opcode | I-type |
| Imm[11:5] | | Rs2 | Rs1 | Funct3 | Imm[4:0] | Opcode | S-type |

Table 2.1.1.15. RV32D RV64D instruction type frames

## RV32D Standard Extension

| Imm[11:0] | | Rs1 | 011 | Rd | 0000111 | FLD |
|---|---|---|---|---|---|---|
| Imm[11:5] | Rs2 | Rs1 | 011 | Imm[4:0] | 0100111 | FSD |
| Rs3 | 01 | Rs2 | Rs1 | Rm | Rd | 1000011 | FMADD.D |
| Rs3 | 01 | Rs2 | Rs1 | Rm | Rd | 1000111 | FMSUB.D |
| Rs3 | 01 | Rs2 | Rs1 | Rm | Rd | 1001011 | FNMSUB.D |
| Rs3 | 01 | Rs2 | Rs1 | Rm | Rd | 1001111 | FNMADD.D |
| 0000001 | | Rs2 | Rs1 | Rm | Rd | 1010011 | FADD.D |
| 0000101 | | Rs2 | Rs1 | Rm | Rd | 1010011 | FSUB.D |
| 0001001 | | Rs2 | Rs1 | Rm | Rd | 1010011 | FMUL.D |
| 0001101 | | Rs2 | Rs1 | Rm | Rd | 1010011 | FDIV.D |
| 0101101 | | 00000 | Rs1 | Rm | Rd | 1010011 | FSQRT.D |
| 0010001 | | Rs2 | Rs1 | 000 | Rd | 1010011 | FSGNJ.D |
| 0010001 | | Rs2 | Rs1 | 001 | Rd | 1010011 | FSGNJN.D |
| 0010001 | | Rs2 | Rs1 | 010 | Rd | 1010011 | FSGNJX.D |
| 0010101 | | Rs2 | Rs1 | 000 | Rd | 1010011 | FMIN.D |
| 0010101 | | Rs2 | Rs1 | 001 | Rd | 1010011 | FMAX.D |
| 0100000 | | 00001 | Rs1 | Rm | Rd | 1010011 | FCVT.S.D |
| 0100001 | | 00000 | Rs1 | Rm | Rd | 1010011 | FCVT.D.S |
| 1010001 | | Rs2 | Rs1 | 010 | Rd | 1010011 | FEQ.D |
| 1010001 | | Rs2 | Rs1 | 001 | Rd | 1010011 | FLT.D |
| 1010001 | | Rs2 | Rs1 | 000 | Rd | 1010011 | FLE.D |
| 1110001 | | 00000 | Rs1 | 001 | Rd | 1010011 | FCLASS.D |
| 1100001 | | 00000 | Rs1 | Rm | Rd | 1010011 | FCVT.W.D |
| 1100001 | | 00001 | Rs1 | Rm | Rd | 1010011 | FCVT.WU.D |
| 1101001 | | 00000 | Rs1 | Rm | Rd | 1010011 | FCVT.D.W |
| 1101001 | | 00001 | Rs1 | Rm | Rd | 1010011 | FCVT.D.WU |

Table 2.1.1.16. RV32D instruction frames

## RV64D Standard Extension (in addition to RV32D)

| 1100001 | 00010 | Rs1 | Rm | Rd | 1010011 | FCVT.L.D |
|---|---|---|---|---|---|---|
| 1100001 | 00011 | Rs1 | Rm | Rd | 1010011 | FCVT.LU.D |
| 1110001 | 00000 | Rs1 | 000 | Rd | 1010011 | FMV.X.D |
| 1101001 | 00010 | Rs1 | Rm | Rd | 1010011 | FCVT.D.L |
| 1101001 | 00011 | Rs1 | Rm | Rd | 1010011 | FCVT.D.LU |
| 1111001 | 00000 | Rs1 | 000 | rd | 1010011 | FMV.D.X |

Table 2.1.1.17. RV64D instruction frames

| λ32 λ28 | λ27 λ26 | λ25 λ21 | λ20 λ16 | λ15 λ13 | λ12 λ8 | λ7 λ1 | |
|---------|---------|---------|---------|---------|--------|-------|--------|
| Funct7 | | Rs2 | Rs1 | Funct3 | Rd | Opcode | R-type |
| Rs3 | Funct2 | Rs2 | Rs1 | Funct3 | Rd | Opcode | R4-type |
| Imm[11:0] | | | Rs1 | Funct3 | Rd | Opcode | I-type |
| Imm[11:5] | | Rs2 | Rs1 | Funct3 | Imm[4:0] | Opcode | S-type |

Table 2.1.1.18. RV32Q/RV64Q instruction type frames

# RV32Q Standard Extension

| | | | | | | | |
|---------|---------|---------|---------|---------|--------|----------|--------|
| Imm[11:0] | | | Rs1 | 100 | Rd | 0000111 | FLQ |
| Imm[11:5] | | Rs2 | Rs1 | 100 | Imm[4:0] | 0100111 | FSQ |
| Rs3 | 11 | Rs2 | Rs1 | Rm | Rd | 1000011 | FMADD.Q |
| Rs3 | 11 | Rs2 | Rs1 | Rm | Rd | 1000111 | FMSUB.Q |
| Rs3 | 11 | Rs2 | Rs1 | Rm | Rd | 1001011 | FNMSUB.Q |
| Rs3 | 11 | Rs2 | Rs1 | Rm | Rd | 1001111 | FNMADD.Q |
| 0000011 | | Rs2 | Rs1 | Rm | Rd | 1010011 | FADD.Q |
| 0000111 | | Rs2 | Rs1 | Rm | Rd | 1010011 | FSUB.Q |
| 0001011 | | Rs2 | Rs1 | Rm | Rd | 1010011 | FMUL.Q |
| 0001111 | | Rs2 | Rs1 | Rm | Rd | 1010011 | FDIV.Q |
| 0101111 | | 00000 | Rs1 | Rm | Rd | 1010011 | FSQRT.Q |
| 0010011 | | Rs2 | Rs1 | 000 | Rd | 1010011 | FSGNJ.Q |
| 0010011 | | Rs2 | Rs1 | 001 | Rd | 1010011 | FSGNJN.Q |
| 0010011 | | Rs2 | Rs1 | 010 | Rd | 1010011 | FSGNJX.Q |
| 0010111 | | Rs2 | Rs1 | 000 | Rd | 1010011 | FMIN.Q |
| 0010111 | | Rs2 | Rs1 | 001 | Rd | 1010011 | FMAX.Q |
| 0100000 | | 00011 | Rs1 | Rm | Rd | 1010011 | FCVT.S.Q |
| 0100011 | | 00000 | Rs1 | Rm | Rd | 1010011 | FCVT.Q.S |
| 0100001 | | 00011 | Rs1 | Rm | Rd | 1010011 | FCVT.D.Q |
| 0100011 | | 00001 | Rs1 | Rm | Rd | 1010011 | FCVT.Q.D |
| 1010011 | | Rs2 | Rs1 | 010 | Rd | 1010011 | FEQ.Q |
| 1010011 | | Rs2 | Rs1 | 001 | Rd | 1010011 | FLT.Q |
| 1010011 | | Rs2 | Rs1 | 000 | Rd | 1010011 | FLE.Q |
| 1110011 | | 00000 | Rs1 | 001 | Rd | 1010011 | FCLASS.Q |
| 1100011 | | 00000 | Rs1 | Rm | Rd | 1010011 | FCVT.W.Q |
| 1100011 | | 00001 | Rs1 | Rm | Rd | 1010011 | FCVT.WU.Q |
| 1101011 | | 00000 | Rs1 | Rm | Rd | 1010011 | FCVT.Q.W |
| 1101011 | | 00001 | Rs1 | Rm | Rd | 1010011 | FCVT.Q.WU |

Table 2.1.1.19. RV32Q instruction frames

## RV64Q Standard Extension (in addition to RV32Q)

| 1100011 | 00010 | Rs1 | Rm | Rd | 1010011 | FCVT.L.Q |
| 1100011 | 00011 | Rs1 | Rm | Rd | 1010011 | FCVT.LU.Q |
| 1101011 | 00010 | Rs1 | Rm | Rd | 1010011 | FCVT.Q.L |
| 1101011 | 00011 | Rs1 | Rm | Rd | 1010011 | FCVT.Q.LU |

Table 2.1.1.20. RV64Q instruction frames

# 2.1.2. Theory and Notes

The RISC-V instruction set listing tables show the frames and instruction set for a particular extension along with the opcode and the mapping of the bit slots to a wavelength for optical computing, with λ1 being the least significant bit slot and λ32 the most significant bit slot. The opcodes in each table can easily be seen to map to a wavelength with an intensity level modulated on to it.

For this project, which is ultimately an attempt at a requirements specification for a holodeck the addressable space needed is in the magnitude of $2^{128}$. Thus the RV32Q and RV64Q standard extensions are needed. For this to be possible, the registers in the RISC-V specification need to be 128 bits wide.

My way of dealing with this is to have loading to and from memory done by the load/store unit. The load/store unit can interface with the address bus, which is 128 bits wide. It can then feed the other units with information. I have allocated a certain amount of memory from this global memory to each thread, and this memory will be distributed with the thread. The grid unit could have its set of memory, which could be the thread's global memory. This would stop an address bus bottleneck.

The integer unit has 32-bit-wide registers.

| Register | Name | Use |
| --- | --- | --- |
| X0 | Zero | hard-wired zero |
| X1 | Ra | return address |
| X2 | Sp | stack pointer |
| X3 | Gp | global pointer |
| X4 | Tp | thread pointer |
| X5 | T0 | temporary/alternate link address |
| X6-7 | T1-2 | temps |
| X8 | S0/fp | saved register/frame pointer |
| X9 | S1 | saved register |
| X10-11 | A0-1 | function arguments/return values |
| X12-17 | A2-7 | function arguments |
| X18-27 | S2-11 | saved registers |
| X28-31 | T3-6 | temps |

Table.2.1.2.1. Register, width 32 bits

The floating point unit has 64 bit wide registers.

| Register | Name | Use |
|----------|------|-----|
| F0->f7 | Ft0->ft7 | Fp temps |
| F8->f9 | Fs0->fs1 | Fp saved registers |
| F10->f11 | Fa0->fa1 | Fp function arguments/return values |
| F12->f17 | Fa2->fa7 | Fp function arguments |
| F18->f27 | Fs2->fs11 | Fp saved registers |
| F28->f31 | Ft8->ft11 | FP temps |
| PC | pc | Program counter |

Table 2.1.2.2. Registers with width 64 bits

In Chapter 13 of the specification, the Q-standard extension for quad-precision floating point version 2.2 is presented, which is to be compliant with the IEEE 754-2008 arithmetic standard.

I am now going to show how the hardware described in Part 1 can be used as a base with slight modifications to other functionality with slight variations. Thus, this maps hardware to instructions. I use tree diagrams in order to show the hardware and then the possible instructions with slight modification.

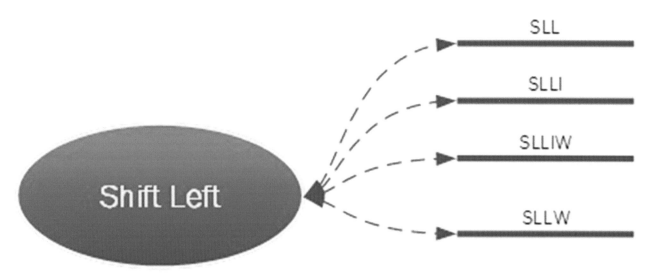

Fig. 2.1.2.3. Shift left

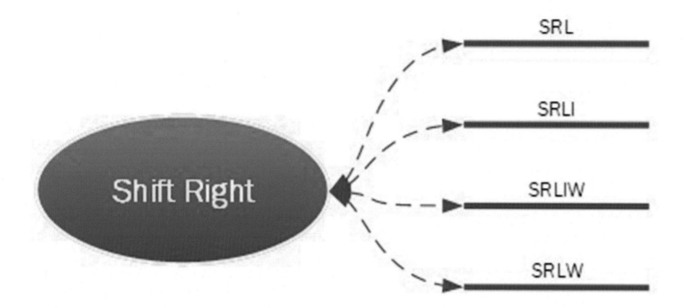

Fig. 2.1.2.4. Shift right

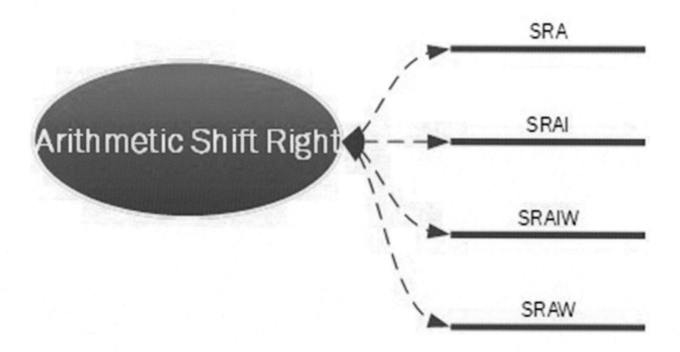

Fig. 2.1.2.5. Arithmetic shift right

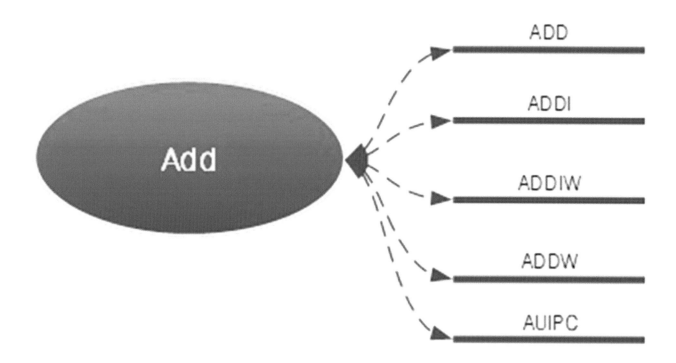

Fig. 2.1.2.6. Add

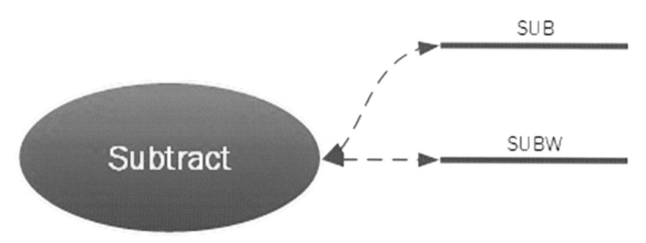

Fig. 2.1.2.7. Subtract

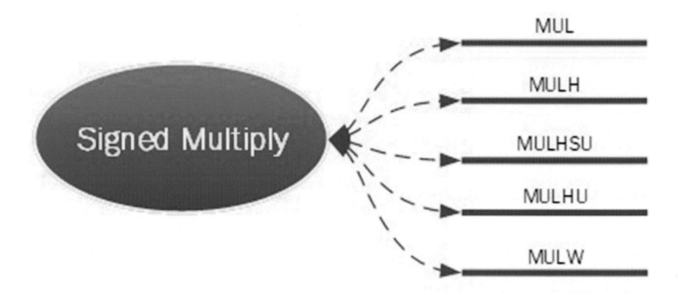

Fig. 2.1.2.8. Signed multiply

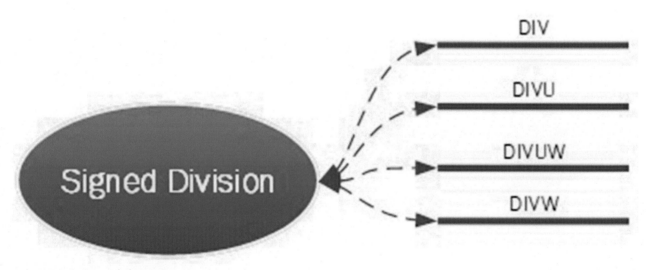

Fig. 2.1.2.9. Signed division

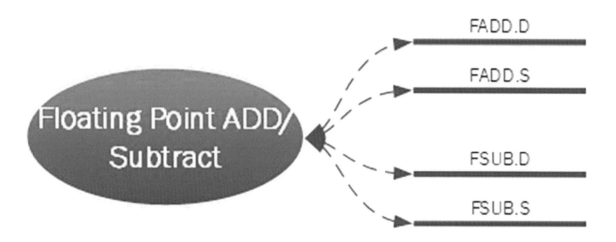

Fig. 2.1.2.10. Floating point add/subtract

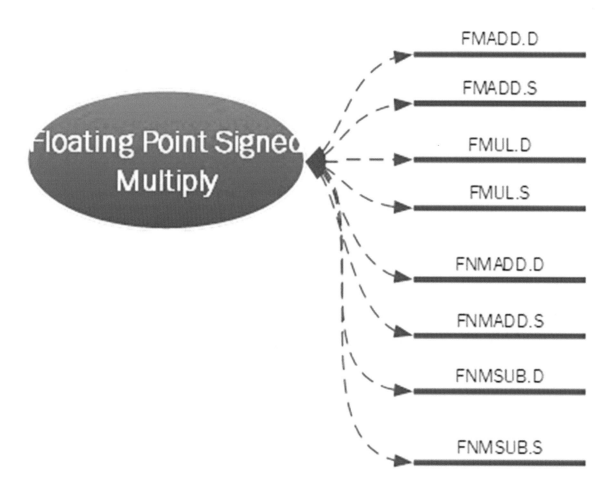

Fig. 2.1.2.11. Floating point signed multiply

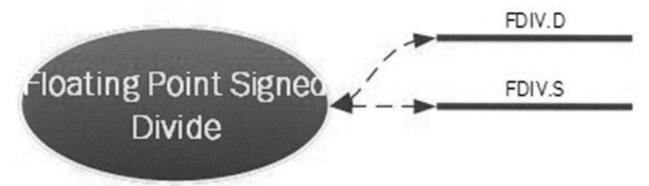

Fig. 2.1.2.12. Floating point signed divide

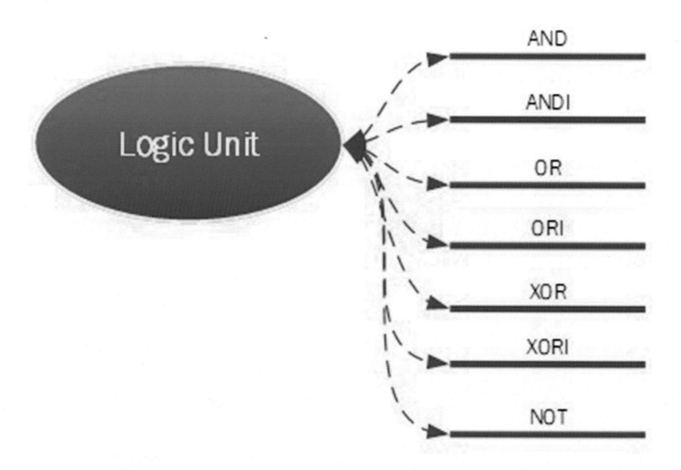

Fig. 2.1.2.13. Logic unit

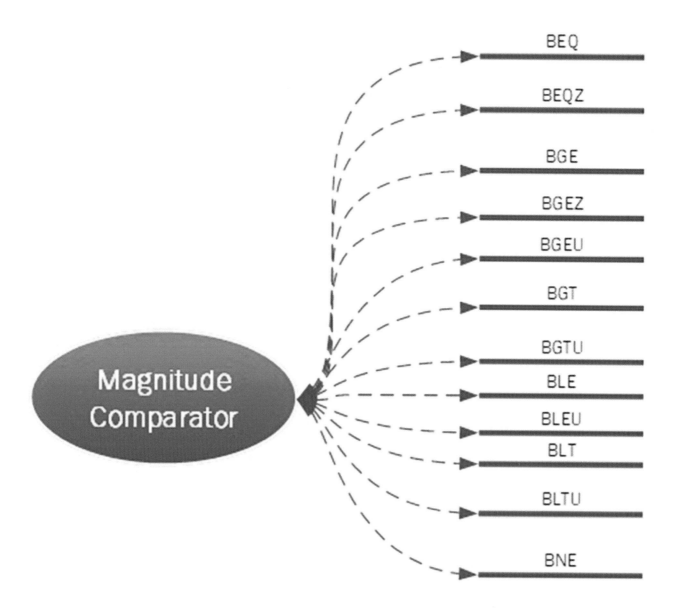

Fig. 2.1.2.14. Magnitude comparator

The magnitude comparator modified for floating point instructions could be capable of determining FEQ.D, FLT.D, FLE.D.

Then, conversion routines may be needed FCVT.S.D, FCVT.D.S, FCVT.W.D, FCVT.WU.D, FCVT.D.W, FCVT.D.WU.

Also FSQRT.D needs to be hardware implemented.

That means with the above said, most of RV32I base instruction set, RV64I base instruction set, and RV32D standard extension can be done where a load/store unit would take care of the loading and storing of values from memory.

This means that several RISC-V instructions will not be implemented, so the implementation will not be wholly RISC-V compliant, but a holodeck is a niche technology with specialist needs.

There is a need to implement 3D math functions.

Cosine and sine (64-bit) are needed especially for the rotation of a voxel functionality, as every voxel has force points within it which are rasterised at 10 per axis (making 1000 per voxel). This means that a voxel length of 20 μm divided by 10 is 2 μm. This means that every force point has the ability to move in any axis x, y, or z at any angle and position within a 2 μm radius of its default position. I am trying to get to the point here that a force point triangulated in sets of three cause a triangle face, which is part of the makeup of the object. Thus a rotate has to be able to deal with the magnitudes of units used in rotations in order to precisely rotate a moment in 3D space.

Vectors are used quite often in 3D calculations. Thus vec2, vec3, vec4, or bool vec2, vec3, vec4 could be hardware implemented, or a script could be written in order to act as part of a larger kernel for each thread. Mat2, mat3, and mat4 could also be implemented in hardware or scripted.

I am now going to briefly describe the material, shading, and lighting models just to give you an overview of what's going on in the data centre virtual node and the personal physical volume nodes for multi-user usage.

Material Theory

In a simulation a material is chosen for a particular object. The material chosen determines the positioning of the moments in order. For example, if a ray of light hits the surface, it is reflected at the colour of the material. Some materials like wood need a noise/pattern in order to show the grain pattern of the wood. This simply adjusts the moments at grain positions to reflect the grain colour.

Shadowing Theory

If a material is solid and all its sides are set to solid, then the rays of light around the object reflect and cause natural shadowing (like in real life).

Lighting Model

If a ray of light hits the face of a moment triangle, it is reflected due to the positioning of the moments. Its colour is also determined by the positioning of the moments. In order to simulate the multi-user holodecks the data centre node has to ray-trace light rays to determine collisions with a face of a moment triangle and determine, due to the moments, the rays' new direction and colour. This is then sent to the Ppv, and there the scene is set up to resemble what is seen virtually on the data centre node.

# Notes

# Part 3

# 3. Volumetric Theory of Operation.

3.1. Introduction

3.2. Basic Volumetrics

3.3. Advanced Volumetrics

3.4. Holodeck Volumetrics

3.5. Holodeck Lighting Theory

3.6. Holodeck Solidness Theory

# 3.1. Introduction

In this section, the theory behind static-volume volumetrics is introduced in a fashion independent of implementation. Then this theory is expanded upon to an advanced level. After this holodeck volumetrics is introduced, and this theory is then shown to use existing light-based voxels as a lighting model for a holodeck. This is then further expanded upon to show several methods at attempts to achieve solidness to touch for projected objects. These are the fundamentals which will be later developed in Part 4 in order to show the inner workings of a holodeck.

# 3.2. Basic Volumetrics

A static volume volumetric display is a display device in which a volume of space is divided up into small volumes called voxels Fig. 3.2.1. These voxels are lit up at various colours depending on the simulation. Where a colour is a wavelength of light in the visible spectrum. Fig. 3.2.2 shows an emitter projecting a voxel into 3D space. The emitter here can be a spatially emitting light device only, which has to be matched to a single point in 3D space. The voxels with this device projecting them are lit at a frequency in order to display to the human eye that a particular voxel is active.

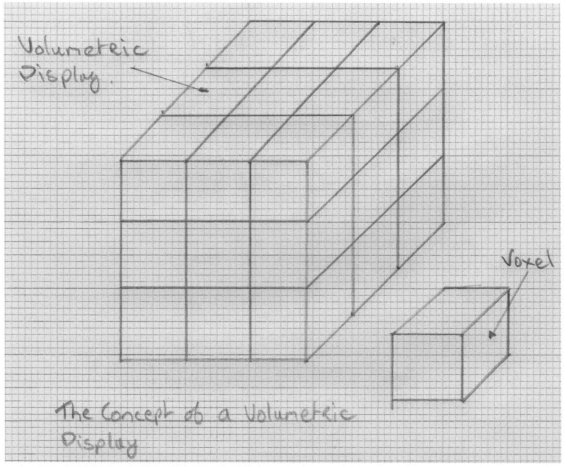

Fig. 3.2.1. The concept of a volumetric display

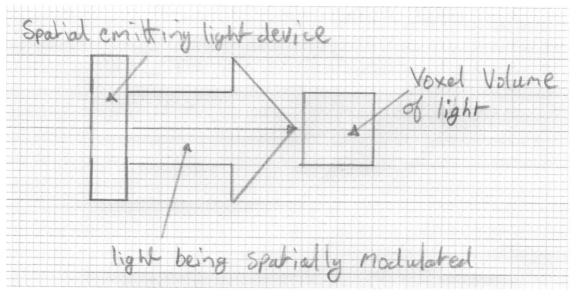

Fig. 3.2.2. The concept of a spatially modulated light wave

It is important to note that light has a wave-like nature and can be expressed in terms of a vector with a direction. So putting this knowledge to use, the light from the spatial light-emitting device is matched to a particular voxel in 3D space. Once this volume is lit by the emitter's projection, the voxel acts like a small light with a view cone in the direction of the wave vector. A view cone in the direction of the vector of the emitted waveform can be seen in Fig. 3.2.3. If the viewer is outside this view cone, then the viewer may not see the voxel of light.

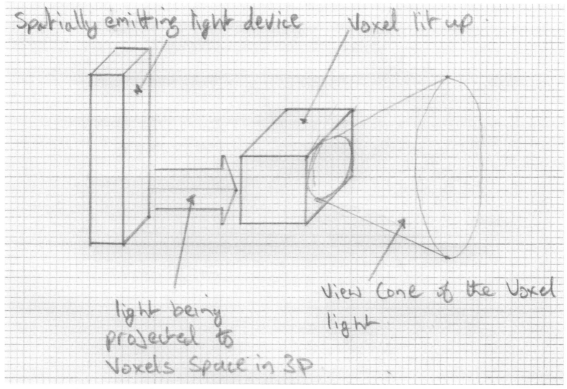

Fig. 3.2.3. The concept of a light wave spatially modulated to a point, and that point being a voxel which has a view cone in the direction of the light wave

# 3.3. Advanced Volumetrics

In Fig. 3.2.3, it shows a limitation to light-based volumetrics as the view cone is limited. A way of overcoming this shortfall is to have several emitters from several directions projecting a voxel into the same place in 3D. In Fig. 3.3.1, it shows this concept with the result of a wider view capability.

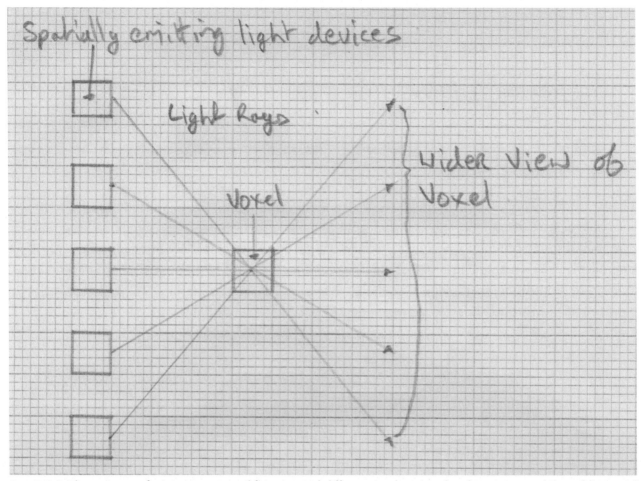

Fig. 3.3.1. The concept of projecting a voxel from several different angles to widen the view capability of the voxel

# 3.4. Holodeck Volumetrics

With metamaterials, it is possible to bend light via negative refraction but the direction vector of the light does not change. In this section when I refer to bending light I mean that the direction vector of the light ray actually changes to a new direction even when the field is no longer present the light ray still propagates in the new direction.

In a conventional Volumetric display a voxel is the smallest point within the display. I am now going to introduce you to my approach to holodeck volumetrics where still a voxel is used, but I use a resolution for the voxel width to be 20 micrometers, which happens to be about the width of a human hair. This voxel is then divided into 10 layers, with 10 force points on each axis. That is a total of 1000 force points which make up a voxel. Note that the amount of force points chosen is a test case, and the number can be changed as needed.

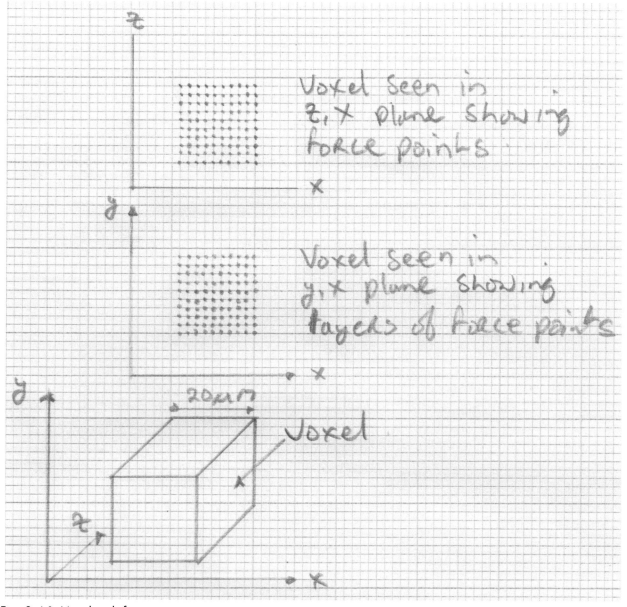

Fig. 3.4.1. Voxel with force points

A force point is a point within a voxel that simulates the magnetic moment of a molecule or atom. So not all force points will be needed for each simulation, just a subset of the 1000 force points. It is important to recognise that the purpose of the force points is to bend a ray of light in order to create a reflection from the surface at a particular wavelength (colour). The colour of the reflected ray can be set by adjusting the width between the magnetic moments and via altering their magnitudes and the moments' x, y, z angles. With this set-up, it may take up to three force points to bend a ray of light. Three magnetic moments on the plane to reflect from interacting with the magnetic moment of the light rays. The three force points on the plane of reflection triangulate the ray of light and can be set up to when the magnetic moment of the ray is reflected. It is then that they set up the direction where the ray of light would reflect. By having these force points on a per voxel basis, a need is there to rasterise the force points into triangles, making a triangle with three magnetic moments, which creates a face of the triangle.

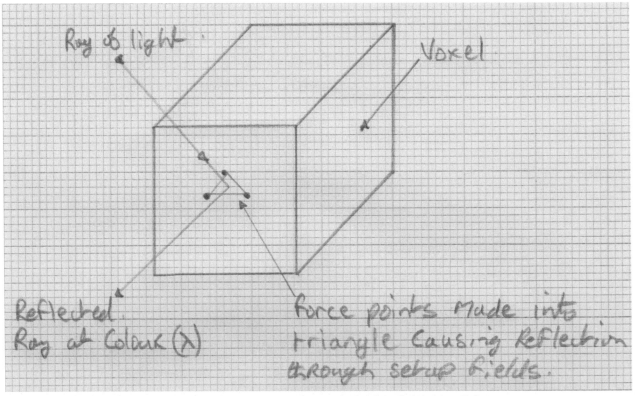

Fig. 3.4.2. Voxel showing three-point moment triangulation. Possible reflection model.

From what has been said so far, it is possible to see that a triangle fan is required in order to create a line within the voxel starting at 3D x, y, z and ending at 3D x, y, z. The width of the line could be set by adjusting the way the rays are repelled from a triangle fan. The width of the triangle fan could be adjusted where every triangle represents a point and, by triangle fanning, the points can link up.

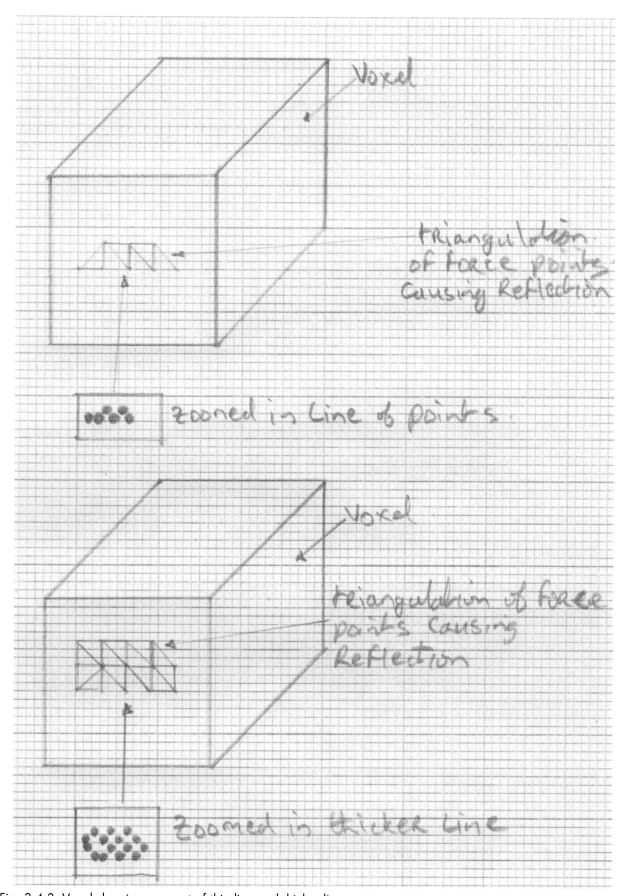

Fig. 3.4.3. Voxel showing concept of thin line and thicker line

**Michael Cloran**

There is a reason I am suggesting that subvoxel force points be used: each ray of light within a simulation that hits the artificial plane set up by the field patterns of the force points have to be reflected, and the rays of light have to always see this plane as there (solid) in order to reflect the rays at the colour of the plane to various set directions which are dependent on the material being simulated. So I don't want some rays being reflected and some not so the update frequency of the force points are set on a per voxel basis, which has to be so that the rays of light always see the artificial plane for reflection.

This means that if a person is within the simulation the artificial material is seen by the observer and if the artificial object is set to solid in all directions (front, back, and both sides) then the person will see natural shadows which are set up by the reflected rays of light which are based on the positioning of the light in the scene.

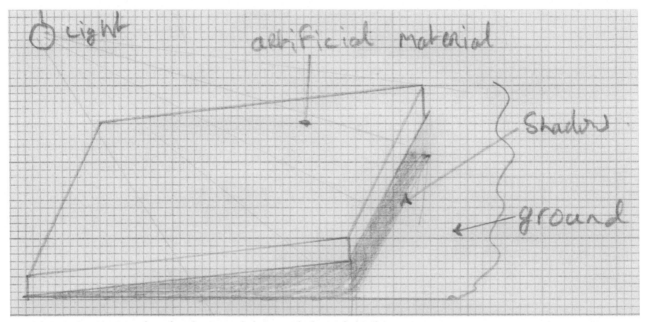

Fig. 3.4.4. Simulation of artificial material showing light and shadow

One problem foreseen with the above theory is that light is made up of components, and when the rays of light are reflected, the other components still propagate at the original direction, so this will upset the simulation as the object will not be solid with these components present. I suggest to either absorb them somehow or to trunk them into a direction such that the observer does not get the other ray components on to their retina for decoding.

94

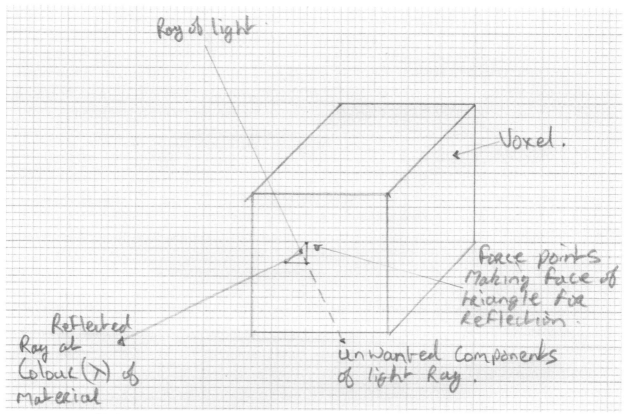

Fig. 3.4.5. Ray of light being reflected at colour but showing unwanted components

# 3.5. Holodeck Lighting Theory

In the basic volumetrics and advanced volumetrics sections, we covered the topic of light-based voxels, where an actual light source projects a voxel into 3D space as a tiny light at a specific colour. Well, with these concepts, instead of reinventing the wheel, why not build on this foundation and use this to light a holodeck scene? The advanced volumetrics section, however, would probably be used more often to build a light source in order to have 360 degrees of lighting on objects within, say, a static scene.

At times, it may not be possible within a scene to use the above method of projecting a light voxel to a 3D point in space because, say, a person gets in the way of the projection device (then the light may appear on them). There are several approaches to creating a light source at a point, one being exciting an electron from one orbit to another for a time. When the electron falls back to its original orbit, it produces a photon. A volume with this approach would produce a voxel of light. With this approach, the lighting modelling as above explained would be the same, just with a different emitter that can project a voxel to a point even when a person is in front of the emitter.

# 3.6. Holodeck Solidness Theory

With superconductors, it is possible to verify that a field pattern can be set up which can be used to levitate certain objects with particular properties. In my research, I have not had the pleasure of having a lab to try out my theories, so I have broken down my approach to solidness into several steps. First, I deal with magnetic moments at the molecular scale, then at the atomic scale, and lastly, with a brute force method. I then suggest using a diamagnetic suit in order to use the property wherein a magnetic moment inside a field is set up in opposition to the fields, thus giving the sensation of touch if the magnitude is right. If all that doesn't work, then I suggest interacting with the inners of an atom in order to get the sensation to touch. This can be done at various levels within the atom. My research is not verified, but I am quite sure that a level of touch can be obtained with one of the following methods. I am now going to explain in more detail.

1. In a simulation, a material is chosen, and the magnetic moments in the artificial material (projected object) is the same as that of the material of choice. As before, the voxels of the material are 20 μm with 1000 force points, and the material being simulated determines which force points are on. The magnitude of the moments would have to be set in order to achieve a level of touch. This is at the molecular scale of simulation.

2. Again, the material of the simulation is chosen, and the force points are set up in order to simulate atomic-level moments in order to get a level of touch, where the magnitude would have to be set certain levels of touch.

3. Brute force. Where all 1000 force points would be on in every voxel taken up by the projected material.

4. To have a tight-fitting diamagnetic suit which would use the property to set up a magnetic moment in opposition to the fields set up by the force points; thus, creating a sensation of touch and solidness to try for 1 to 3 above.

5. To interact with the inners (particles inside) of an atom.

Note that 1 to 4 are electromagnetic.

It should be noted that for safety reasons, bio-interaction simulations and tests should be done prior to testing for touch with a real person.

# Part 4

# Holodeck Part of Book

In this section, there are three chapters which introduce you to the theory of operation of a single-user holodeck, a scrolling holodeck and how a networked and multi-user holodeck could work, and then lastly, some simulation concepts are introduced.

# 4.1. Holodeck Theory of Operation

4.1.0. Architecture Overview

4.1.1. Architecture Mappings from Volumetric Space to Motherboard Thread Space

4.1.2. Scene

4.1.2.1. Basic Scene

4.1.2.2. Light

4.1.2.3. Terrain

4.1.2.4. Sound

4.1.2.5. Touch

4.1.2.6. Different Materials

4.1.2.7. Different Textures

4.1.2.8. Different Colours

4.1.2.9. Resolution of Force Points

4.1.2.9.1. How to Change Update Frequency

4.1.2.10. How a Scene Reacts to Movements of Characters/People

4.1.2.11. Scrolling scene

4.1.2.11.1. Light

4.1.2.11.2. Terrain

4.1.2.11.3. Sound

4.1.2.11.4. Touch

4.1.2.12. Bounds

## 4.1.0. Architecture Overview

The architecture for a holodeck is massively parallel, where massive amounts of data need to be updated at, say, 60Hz (depending on the simulation) at the animation layer. But at the low-level force point update layer (for objects in the scene that are solid or reflect light or display light) the update frequency would need to be independent of the animation frame rate.

## 4.1.1. Architecture Mappings from Volumetric Space to Motherboard Thread Space

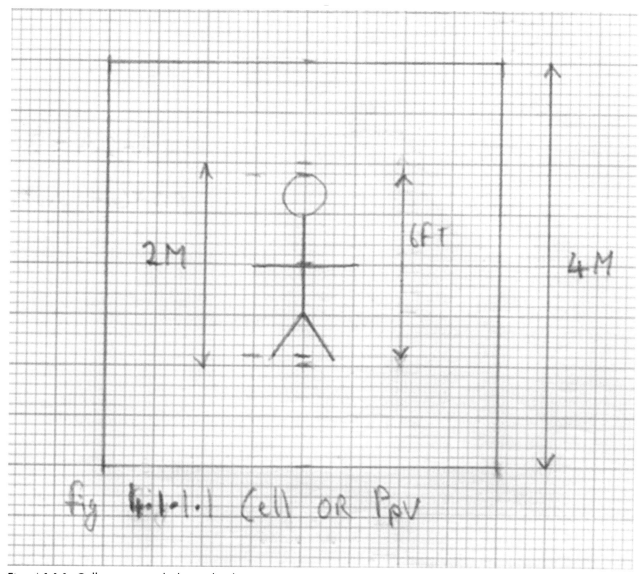

Fig. 4.1.1.1. Cell or personal physical volume

If the volumetric display space is divided into cells where each cell is a personal physical volume (Ppv), for an average man with 6 foot height = 2 × height = 12 feet. Normalised to meters, this is approximately 4 metres, and the volume is 4 m × 4 m × 4 m = 64 m³ (Fig. 4.1.1.1). A volumetric display can be one or more of these cells (Fig. 4.1.1.2).

For a cell/Ppv of 64m³ where the voxel length (Fig. 4.1.1.1A) is 20 μm, there would be

4m/20 μm = 200,000 voxels—that is 200,000 × 200,000 = 40,000,000,000 voxels per light plane.

With 200,000 light planes, that would lead to

200,000 × 200,000 × 200,000 = 8,000,000,000,000,000 voxels—8 petavoxels within the Ppv.

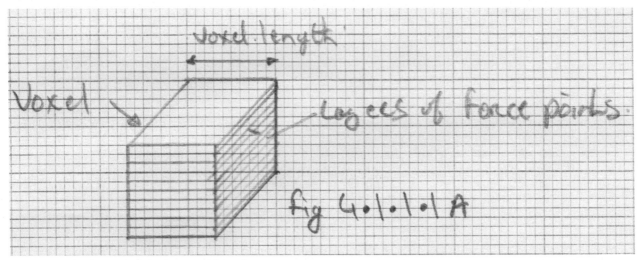

Fig. 4.1.1.1A. A voxel showing voxel length

Force Points

Force points are used to simulate the field pattern of molecular or atomic fields at a point within or on the surface of a material, where the material is the physical material to be simulated.

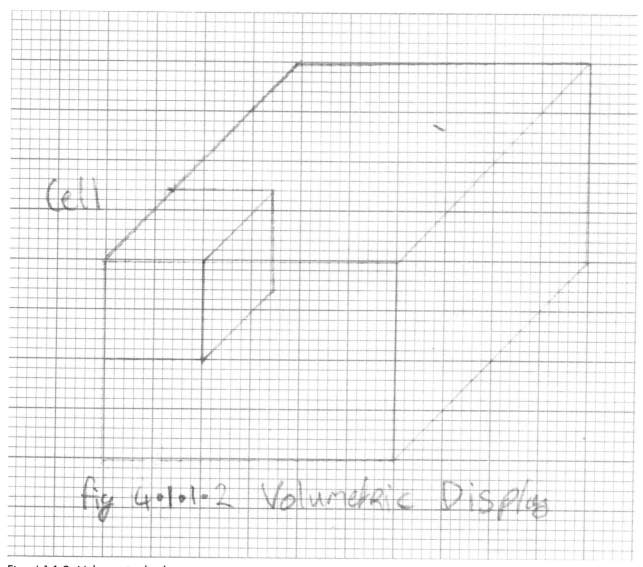

Fig. 4.1.1.2. Volumetric display

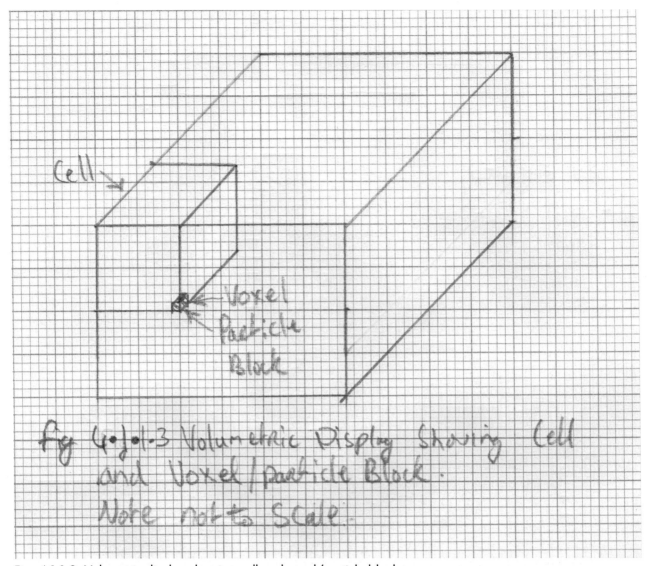

Fig. 4.1.1.3. Volumetric display showing cell and voxel/particle block

If every cell is broken down into subcells and each subcell has a voxel which has a force point block (also described as a particle block) (Fig. 4.1.1.3), then to map from a subcell to a core which contains threads (Fig. 4.1.1.4), a node on the motherboard would correspond to a cell in the volumetric display, which is a personal physical volume. A force point or group of force points can then be easily referenced via cell(x) broken down to subcell, broken down to core, and further broken down to thread of cell(x).

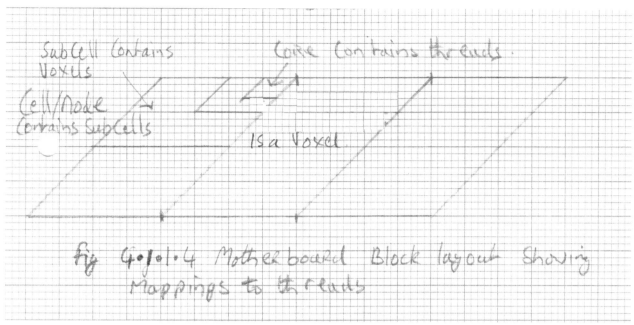

Fig. 4.1.1.4. Motherboard block layout showing mappings to threads

From below, in order to access the node on the motherboard, a method of mapping volumetric space to motherboard node space has to be achieved. So if you want to access the thread that deals with a force point or group of force points (depending on the resolution of threads, one per force point or, say, A force points per thread) within a voxel on cell/Ppv (1) subcell grid unit(2) voxel/core (3) force point layer(1) particle/group(10),

Thread(1, 2, 3, 1, 10)

Node on motherboard = Ppv = CELL(X) in volumetric display.

If you want to reference the subcell(X, I)

If you want to reference the core(X, I, N)

If you want to reference the thread(X, I, N, M)

From above, if a voxel is broken down to, say, 10 layers with 100 force points per layer. If a thread can take care of a group of force points—say, 10—then for each voxel there are 100 threads. Thus, for a Ppv 8,000,000,000,000,000 voxels, there would have to be 800,000,000,000,000,000 threads. If each thread has an amount of RAM available to it—say, if 500 KB is available per thread—then the RAM required would be $4.096 \times 10^{23}$ RAM. This RAM would be shared in buffers. Thus the design would be a shared-memory architecture.

There is a problem with this scenario: $2^{64} = 18,446,744,073,709,551,616$ clearly shows that a 64-bit address bus with binary addressing would not be enough to deal with the addressable space ($2^{128} = 3.4028 \times 10^{38}$ approximately). So the address bus size would have to be greater than 64 bits and less than 128 bits.

# 4.1.2. Scene

A scene in a holodeck is a simulation of a physical environment where every physical object has a certain level of touch based on the simulated material, texture, and colour. The scene is set up in order to portray a real environment.

A scrolling scene is set up to keep the person in the simulation approximately in the centre of the scene's physical space, but the person can be anywhere in the scene's virtual space (Fig. 4.1.2.1).

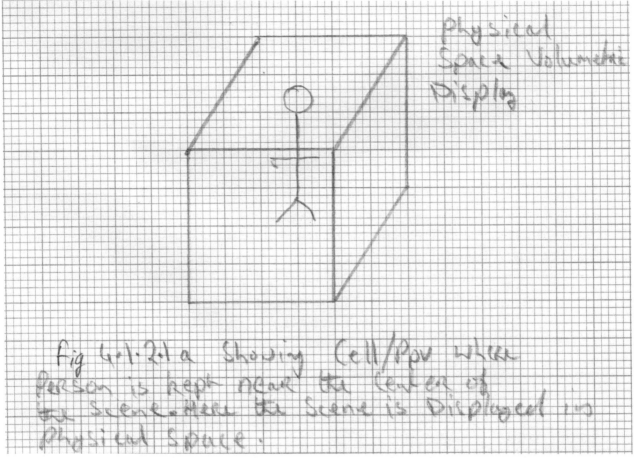

Fig. 4.1.2.1a. Cell/Ppv where person is kept near center of the physical space

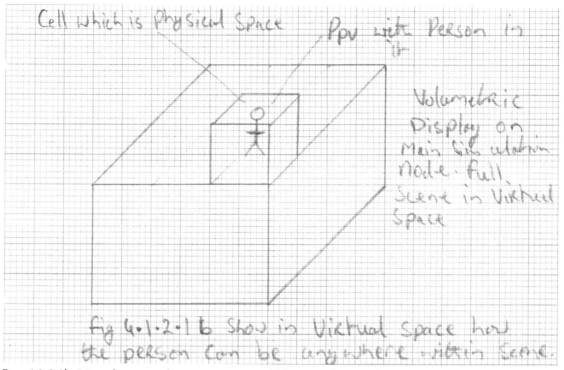

Fig. 4.1.2.1b. Virtual space where a person could be anywhere within scene

## 4.1.2.1. Basic Scene

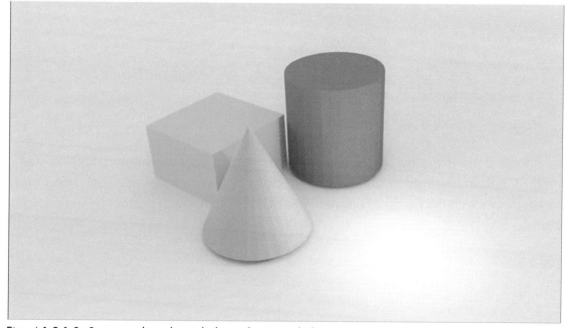

Fig. 4.1.2.1.0. Scene with ambient light and a point light

Fig. 4.1.2.2.0 shows a very basic scene with a world with three objects in it: a box, a cone, and a cylinder. The scene has a ground and a point light. The objects are made up of a material and have texture to touch and have a colour. The ground also has a material and texture to touch and a colour. The point light shines quite bright with a colour. The light itself has a material and texture to touch and a colour (translucent).

## 4.1.2.2. Light

Light

The light in Fig. 4.1.2.1.0 has a lighting model with ambient and a point light present.

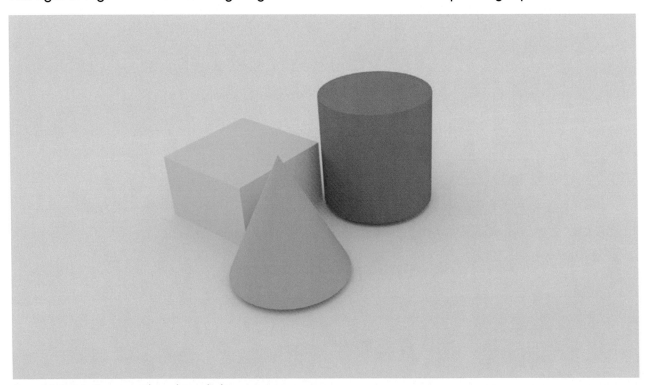

Fig. 4.1.2.2.1. Scene with ambient light

Ambient Light

Ambient light is present via the use of low-light-level spatial light modulators lighting non-object voxels with a certain low-light level. This can be done on every voxel not containing an object, be done randomly with the range of unused voxels, or be scripted to project light along a certain path (Fig. 4.1.2.2.1).

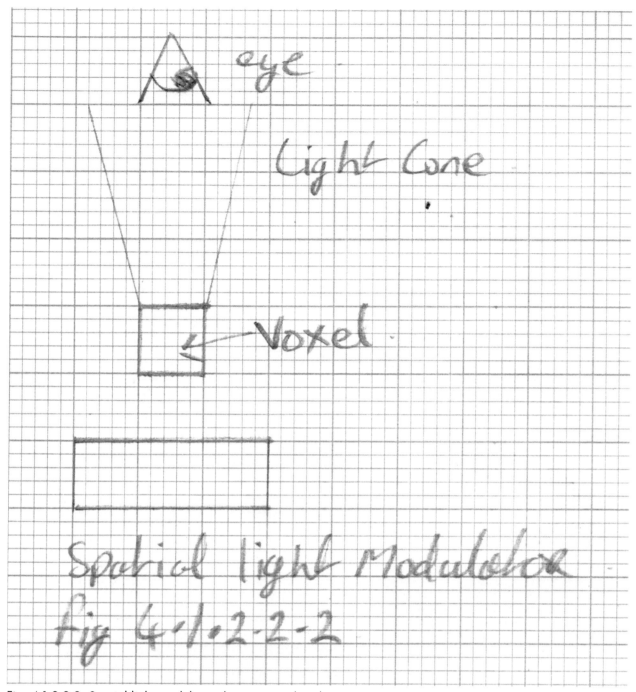

Fig. 4.1.2.2.2. Spatial light modulator showing voxel and view cone

Fig. 4.1.2.2.2. shows light at a point emitted from a spatial light modulator. With this set-up, the light can only be seen by an observer within the light cone of the spatially emitted light. Fig. 4.1.2.2.2 also shows a certain light cone with a viewing angle.

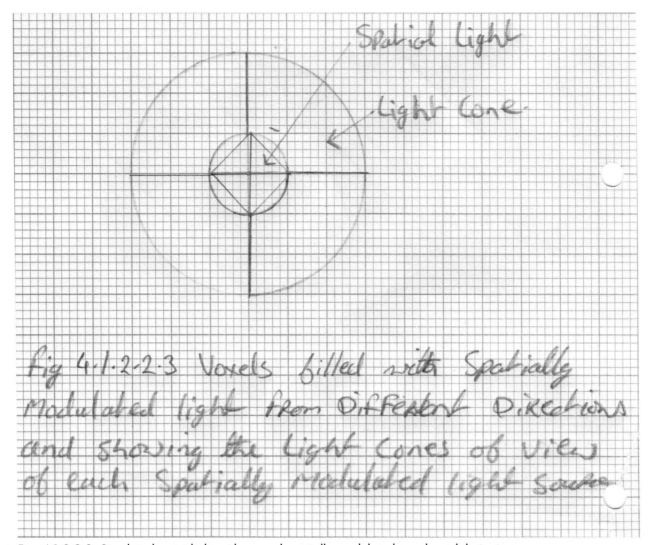

Fig. 4.1.2.2.3. Simulated point light with several spatially modulated voxels and their view cones

## Point Light

For a point light which is omni-directional, many spatial light modulators could be used for a number of degrees' increment (Fig. 4.1.2.2.3), where the degrees of increment would depend on the resolution of viewing light required. A 3D omnidirectional light can be envisioned via setting up spatial light modulators in the three axes (x, y, and z). Just imagine a light is hanging from the ceiling so light has to be shined on the ceiling and in all directions. This set-up for a light bulb would require setting up the light cones so as to portray the view of a real light bulb.

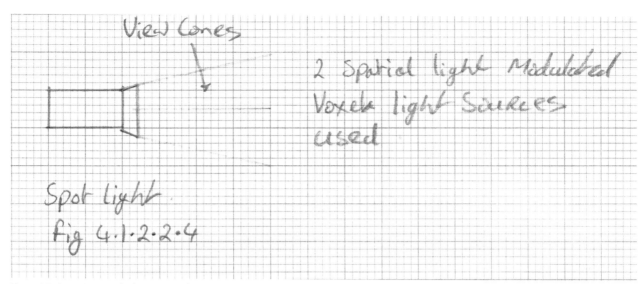

Fig. 4.1.2.2.4. Spotlight in 2D showing view cones

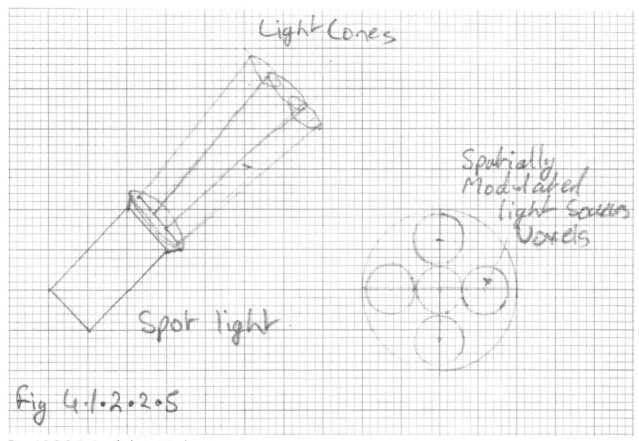

Fig. 4.1.2.2.5. Spotlight in 3D showing view cones

Spot Light

A spot light can be set up via the use of one or many spatial light modulators. Fig. 4.1.2.2.4 shows a flash light with two spatial light modulators making up the x and y axis of the light, but Fig. 4.1.2.2.5 shows a more approximate solution with five spatial light modulators approximating the light cone of a real flash light.

# 4.1.2.3. Terrain

The terrain in Fig. 4.1.2.1.0 is just a flat surface which has a material and texture and a colour.

Material

A material has a certain molecular structure which sets up fields depending on the material in question.

If a volume within the volumetric display is taken to be the voxel in question and a person was to try and touch the ground, they would feel the presence of a material. If this presence or force was sufficient enough, touch could be experienced.

I'll explain. A material is made up of a molecular structure which has a certain signature due to the chemical makeup of the material. If a field pattern was set up with force points at the resolution of the material in question, then touch could be possible either via electromagnetics or by a more complex method (such as to interface with the individual atoms).

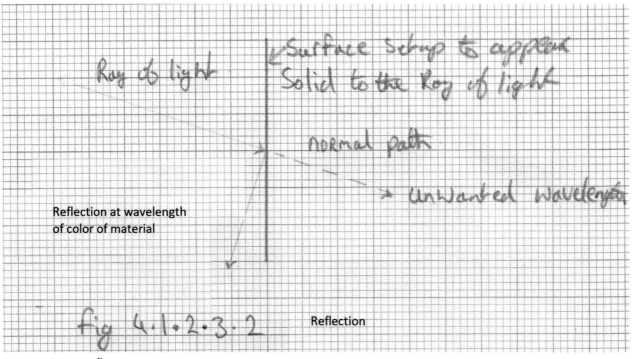

Fig. 4.1.2.3.2 Reflection

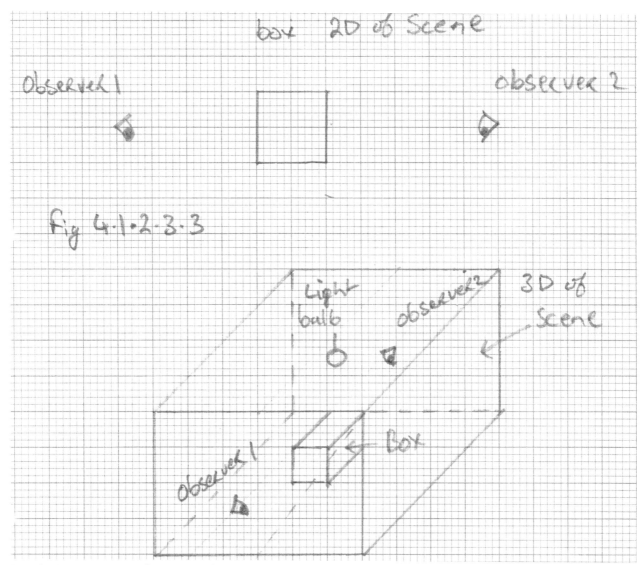

Fig. 4.1.2.3.3. A 2D and 3D view of a basic scene with two observers

If interfacing with the molecules is deemed as enough to give touch, then I would also see the field pattern such that it causes reflection at particular wavelengths in the optical range of vision (causing reflection at a particular colour). For the other optical wavelengths, two options are available to deal with them so as the object appears solid to the eye. The options are (1) to somehow absorb the other optical wavelengths, or (2) to trunk them out in a direction so that observers in the scene don't get the presence of the unwanted rays. Thus with this type of set-up, the material would appear solid to the eye and the eye would see its colour via reflection. And if the force point resolution and strength is sufficient, then touch could be envisioned. Fig. 4.1.2.3.3 shows a room scene with a light and a box with two observers. In this scenario, the light is over the box, so the trunking of the unwanted wavelengths would be downward via bending the light beam with the various wavelengths that are present.

Sidestep

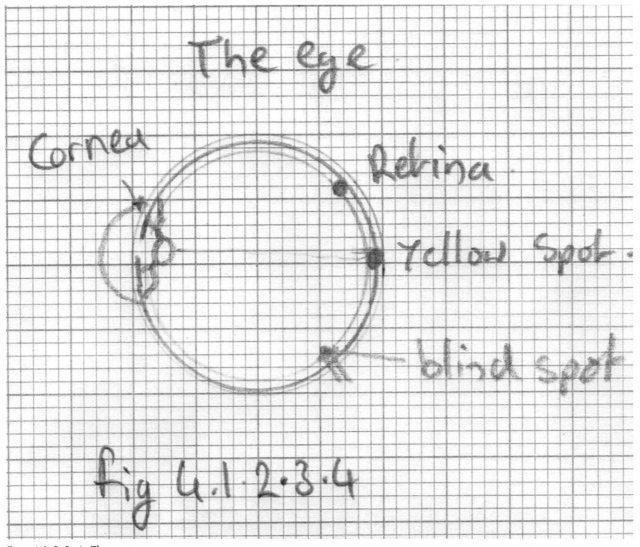

The eye

Cornea

Retina

Yellow Spot

blind spot

fig 4.1.2.3.4

Fig. 4.1.2.3.4. The eye

The operation of the eye is shown in Fig. 4.1.2.3.4.

In order for a human being to see an object, a ray of light has to be reflected from the object and that ray of light has to enter a person's eye and to be present on their retina.

## 4.1.2.4. Sound

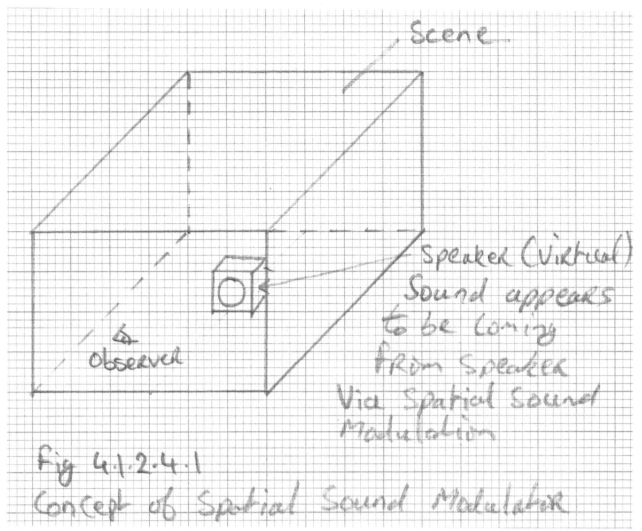

Fig. 4.1.2.4.1. Virtual speaker concept

For sound to work in a holodeck environment, it would have to be able to reproduce sound at points within a scene. For an example, Fig. 4.1.2.4.1 shows a simple scene with a speaker in the room. The environment would have to be able to produce a sound from the virtual speaker. Let's call this *sound at a point*, and let's say a spatial sound modulator was used to at the point of the speaker to have the sound play from that point only. The actual propagating of the sound in this simple scene would be from the point the spatial sound modulator is modulated to.

Definition.

A *spatial sound modulator* is 'a device to create sound at a point in 3D space'.

## 4.1.2.5. Touch

Touch is the ability to touch an object within a scene and be able to physically experience solidness and its texture.

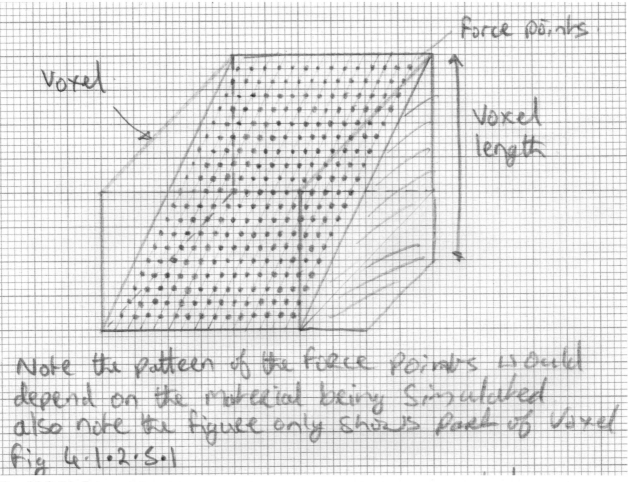

Fig. 4.1.2.5.1. Force points

This could be achieved by artificially creating the field patterns (Fig. 4.1.2.5.1) of the material of the object in question. Let's call the field patterns at points in 3D *space force points*. If these are put in place by a spatial modulator, the update would have to be such that a ray of light sees the force points in the pattern of the material in question, and thus reflection would take place when light shines upon the object (Fig. 4.1.2.5.2) as well as delivering the sensation of touch.

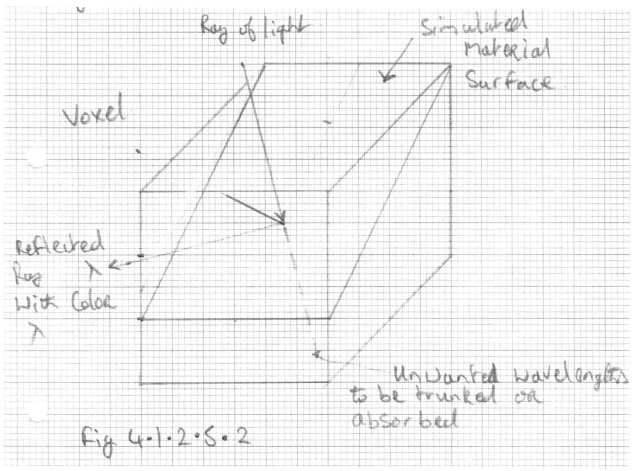

Fig. 4.1.2.5.2. Voxel showing reflection from simulated surface

The force points would have a certain force in newtons of resistance. So by adjusting the resistance force, water and solids could be simulated.

## 4.1.2.6. Different Materials

Materials are made up or one or more molecules. For a holodeck environment, the outer skin would have to be simulated with the molecular structure field pattern of the molecules at the skin. This would have to be made to have the force points (explained above) such that the desired texture is achieved to touch and reflection is achieved. Thus it would be possible to see the colour of the object at the surface level. Fig. 4.1.2.6.1 shows a material made from several chemical compounds, and Fig. 4.1.2.6.2 shows the simulated holodeck version.

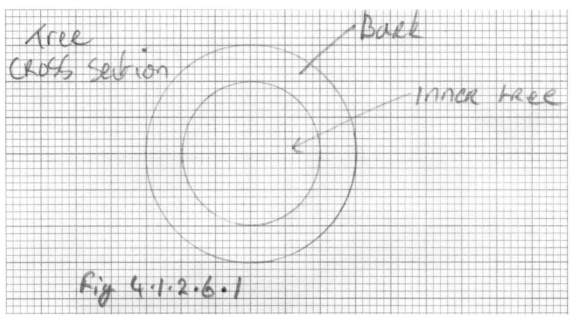

Fig. 4.1.2.6.1. Cross section of tree

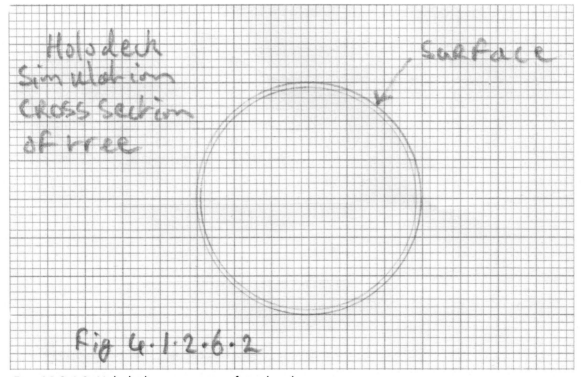

Fig. 4.1.2.6.2. Holodeck cross section of simulated tree

## 4.1.2.7. Different Textures

The texture of an object is basically the feel to touch as well as its appearance. With this definition, if if the skin in a holodeck environment had a thickness and the force points at the skin were set to a certain newton force, then a force that exceeds this would cause penetration of the surface where the penetrating object would feel the resistance force. Fig. 4.1.2.7.1.

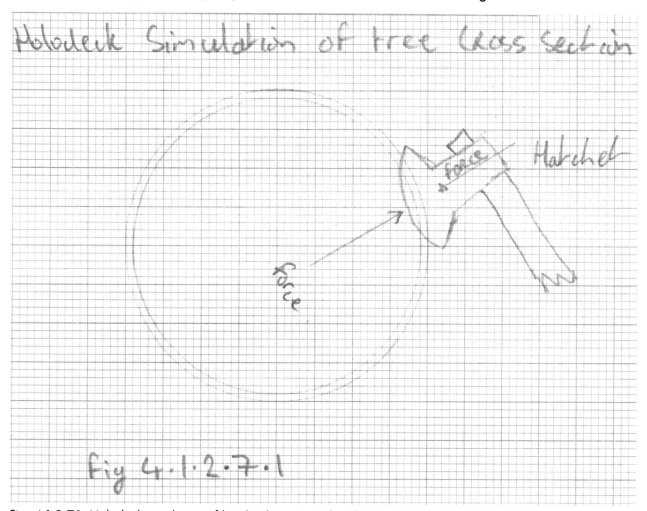

Fig. 4.1.2.7.1. Holodeck simulation of hatchet hitting simulated tree

## 4.1.2.8. Different Colours

For different colours to be present different, materials with textures would have to be simulated at a chemical compound level. Thus at the surface of the material, a field pattern would be present which would reflect the wavelength of the colour of the material (Fig. 4.1.2.3.2, reflection).

Note that in a simulated environment, the colour is reflected, but the other colours in the optical spectrum (Fig. 4.1.2.8.1) have to be dealt with either by somehow absorbing the unwanted colours or by bending them into a kind of trunk to guide them to a direction in order for them not to be seen by observers within the scene (Fig. 4.1.2.3.3).

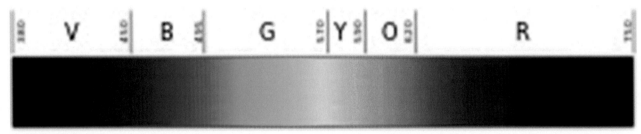

Fig. 4.1.2.8.1. Optical spectrum

Table 4.1.2.8.1. Colour and wavelength table

| Colour | Wavelength |
|--------|------------|
| Violet | 380-450nm |
| Blue | 450-495nm |
| Green | 495-570nm |
| Yellow | 570-590nm |
| Orange | 590-620nm |
| Red | 620-750nm |

# 4.1.2.9. Resolution of Force Points

Every voxel contains an amount of force points which can be turned on or off depending on the material being simulated.

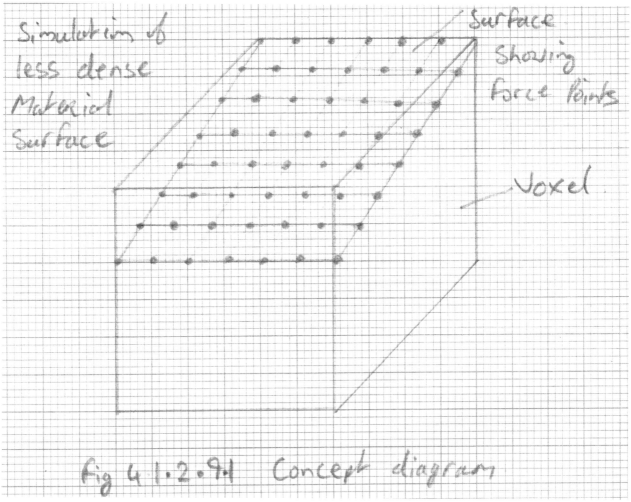

Fig. 4.1.2.9.1. Simulation of less dense material concept diagram

For all molecular structures, the density of the force points has to be able to be set up to simulate the molecules. So for a particular simulation of a less dense molecule (Fig. 4.1.2.9.1) only the density of force points required to simulate at the skin of the surface would be required; the rest of the force points would be turned off at the emitters. However, for a very dense compound, all the force points in the voxel at the skin could be required (Fig. 4.1.2.9.2).

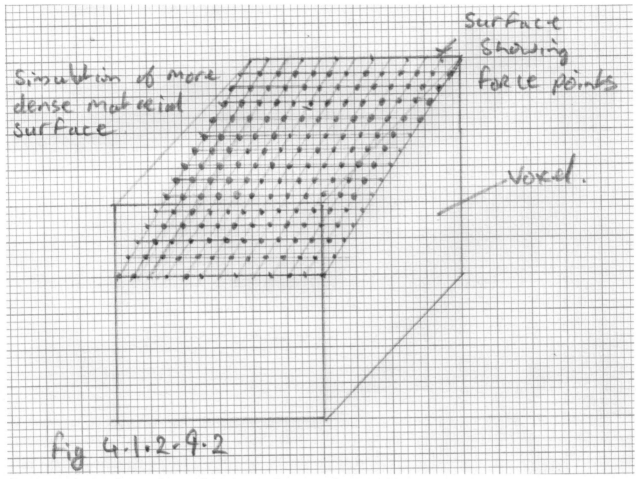

Fig. 4.1.2.9.2. Simulation of more dense material concept diagram

# 4.1.2.9.1. How to Change Update Frequency

For scientific simulation, every voxel would have to have a specific update frequency to be seen as either a light or a solid or undefined.

If a voxel is a light, then a spatial light modulator updates the voxel at a particular frequency—this would be the frequency of the light. If the voxel is a solid, then the update frequency of a spatial modulator has to be such that a ray of light has to see the point it hits within the voxel as solid and reflection occurs (Fig. 4.1.2.9.1.1). Also during a scientific simulation of, say, a force field, in order to reflect certain lasers, the frequency of update would have to be changed to be seen as solid from the laser's perspective.

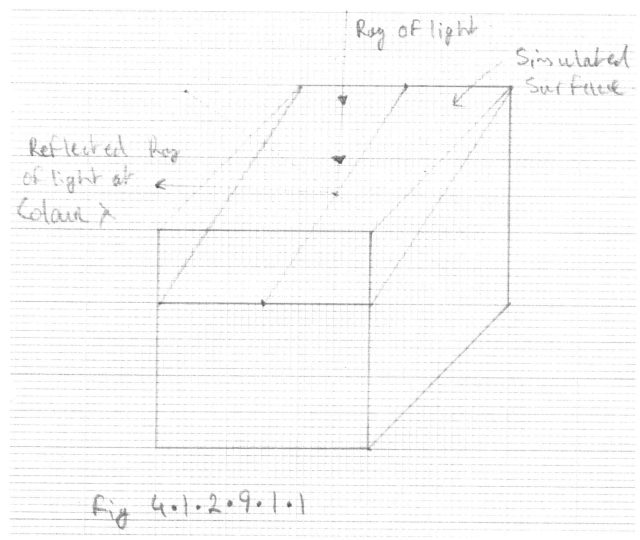

Fig. 4.1.2.9.1.1. Ray of light reflecting within a voxel

## 4.1.2.10. How a Scene Reacts to Movements of Characters/People

In the basic scene (Fig. 4.1.2.1.0), if a person was to walk around it, the shadows would be natural depending on the position of the person and the light bulb. In relation to the terrain, the floor panels have a material with a texture which is set to a newton force. If this is exceeded, then the person would fall through the terrain but experience a resistance.

Example: A person weights 76 kg, and the gravitational pull of planet earth is 9.8m/s$^2$. Then the newton force required to hold the person above the terrain would be 76 × 9.8 = 744.8 newtons.

Thus the force points would have to be able to hold this force no matter what the surface area of the person is exposed to the terrain.

For instance, if the person stands on their toes on one leg, then only the surface area of the toes are hitting the terrain. But still, this surface area feels the full downward force of 744.8 newtons and has to resist this force.

## 4.1.2.11. Scrolling Scene

For a holodeck with a scrolling scene the concepts of operation are more complex than the simple scene in Fig. 4.1.2.1.0. The full scene has to be loaded into memory, and the scene is scrolled based on the physical space available (Fig. 4.1.2.11.1). So the main simulation node would need more storage capacity and processing power than for the scene in Fig. 4.1.2.1.0.

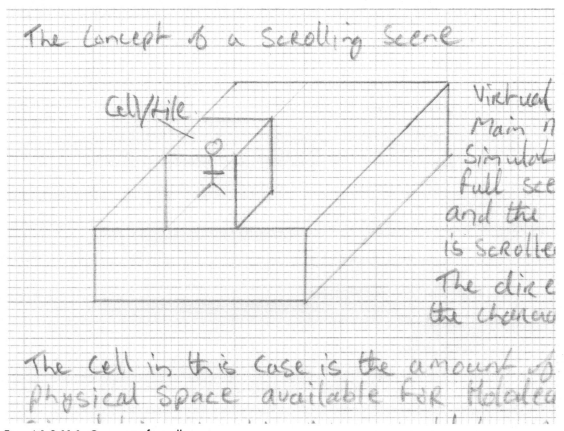

Fig. 4.1.2.11.1. Concept of scrolling scene

Operation

When a person walks in a scrolling scene, the world moves around them, and scrolls in the direction of the walking person.

Limitation

For this type of scrolling scene, the physical space limits the user interaction in that if more than one person was in the scene and they walked in different directions, then it would not work. I will deal with multi-user holodecks later with person physical volumes.

# 4.1.2.11.1. Light

Light in a scrolling scene has to be ray-traced on the simulation main node with the virtual scene and recreated in the physical space holodeck room. Via ray-tracing the light, light levels can be calculated at the outer entry points to the physical space. And by placing strategic spatially modulated light points, the light in the physical space for basic scenes can, in theory, be as in the real world (Fig. 4.1.2.11.1.1). Note that many light sources may be present, and some will be outside the current tile space but will cause lighting effects within the tile.

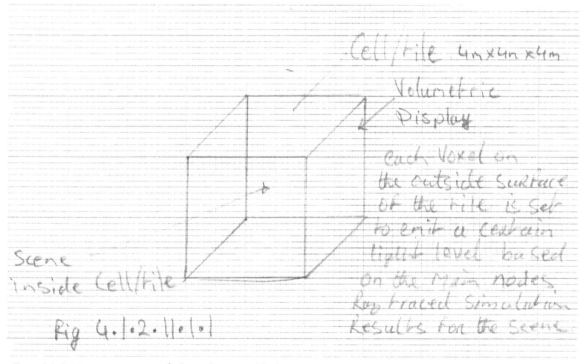

Fig. 4.1.2.11.1.1. Concept of tile with spatial light modulator spatially modulating to the outer surface of the tile/cell in order to create the scene with these light points per voxel on the surface creating the light for the scene

Definition

A tile has a volume which is the physical space available to the system. Thus, the system scrolls volume by volume in a increment in the direction of the character.

## 4.1.2.11.2. Terrain

The terrain in a scrolling scene is tiled based on physical space; thus, when the character walks, the scroll is in the direction of the character. It is important to visualise the scene moving around the person in the scene. The person would be nearly in the same position in physical space with slight deviations most of the time.

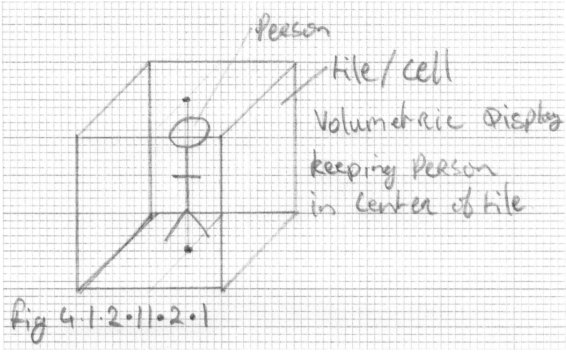

Fig. 4.1.2.11.2.1. Concept of centering the user within the tile

The terrain in the scene can be made up of several different materials in order to portray a solid ground with, maybe, a muddy surface with water over it. For example, walking through a shallow river or walking on the muddy banks of a river. The terrain can also be varied in that it can have mountains and cliffs, and in a simulation, a person could climb a mountain or walk up a cliff (Fig. 4.1.2.11.2.2, Fig. 4.1.2.11.2.3).

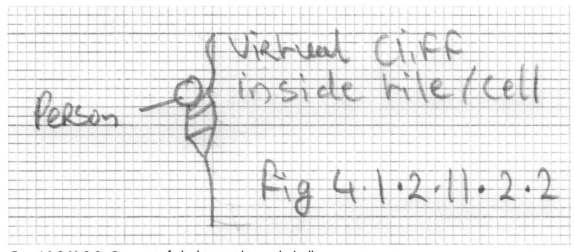

Fig. 4.1.2.11.2.2. Concept of climbing within a tile/cell

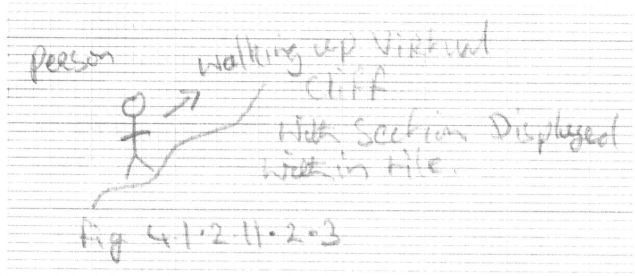

Fig. 4.1.2.11.2.3. Concept of walking up a cliff within a tile/cell

## 4.1.2.11.3. Sound

Sound in a scrollable scene has to be ray-traced and recreated via spatial sound modulators. The sound source is simulated on the main simulation node and based on this simulation the spatial sound modulators set up sound source sound points for the physical space and the person in the scene hears the sound in a spatial context and at the correct amplitude based on the virtual distance from the sound source to the person. Fig. 4.1.2.11.3.1.

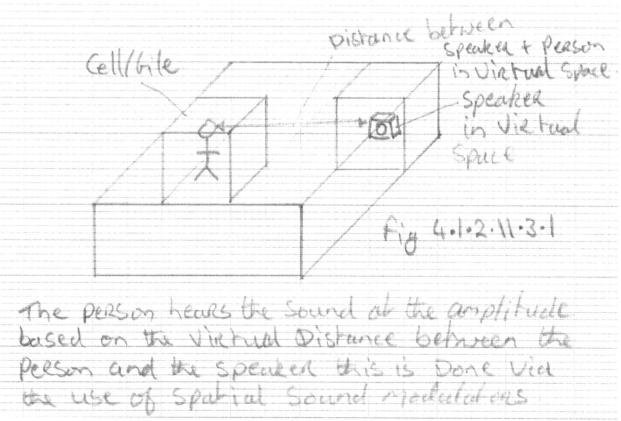

4.1.2.11.3.1. Concept of virtual sound from within scene

## 4.1.2.11.4. Touch

For touch to work in a scrollable scene based on the tile the person is in, if touch points are set (say, walls or ground or objects within the scene are present) and they are meant to be touchable, then the force points explained before are in play and are set up on a per voxel basis to simulate the material of the object at the surface. Then, based on a newton force set by the simulation designer, the touchable objects can be touched, and the objects will feel like the texture of the material of the object. Water can be imagined by setting a low newton force of resistance to touch (Fig. 4.1.2.11.4.1, wall); Fig. 4.1.2.11.4.2, water).

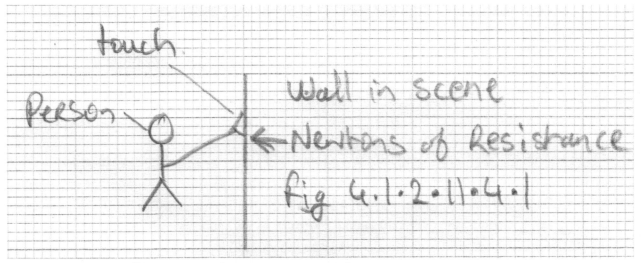

Fig. 4.1.2.11.4.1. A wall in a scene with a newton resistance force

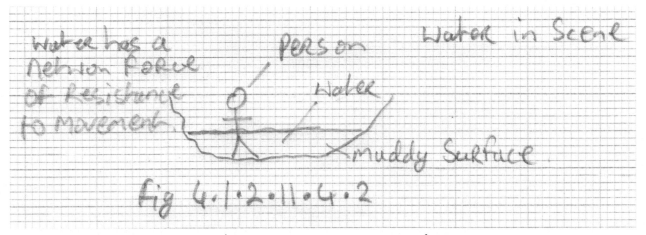

Fig. 4.1.2.11.4.2. Water in a scene with a newton resistance to movement force

# 4.1.2.12. Bounds

The bounds of an object is the surface of the object. If the object is solid, the shape of the object is determined on the amount of force points used per voxel. For instance, the object could only cut half the voxel for solid; thus, half the emitters would be turned on for the voxel in question. This is achieved by having a bit buffer for each force point, and if set, the force point is on. Thus a rasterisation unit is required per voxel for fine textures and shapes. This will allow light-level detail and minute fine materials or even several materials with intertwined textures.

The bounds will have a set level of resistance to touch force. If this is exceeded, then the bounds will be penetrated. A sensor with fine detail is needed to detect the bounds penetration, and thus in the simulation, this can trigger an event. For example, let's consider a hatchet cutting a piece of wood. The bounds of a piece of wood is penetrated, and based on a simulated outcome, the piece of wood is split in half (Fig. 4.1.2.12.1, Fig. 4.1.2.12.2).

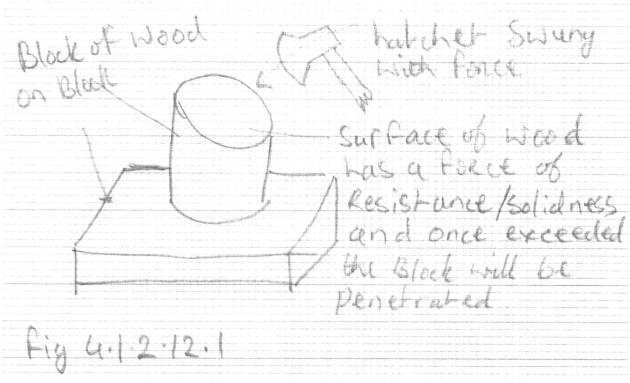

Fig. 4.1.2.12.1. Hatchet being swung to penetrate a block of wood

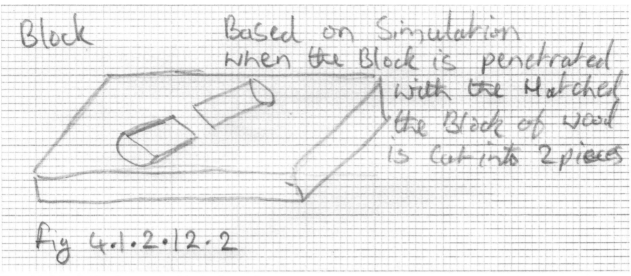

Block

Based on Simulation
when the Block is penetrated
with the Hatchet
the Block of wood
is Cut into 2 pieces

fig 4.1.2.12.2

Fig. 4.1.2.12.2. Hatchet cut the block of wood into two pieces

# 4.2. Networked and Multi-User Holodeck

4.2.1. Introduction
4.2.2. Architecture Overview
4.2.3. Scene on Main Node
4.2.4. Multiple Ppv Nodes
4.2.5. Ppv Physical Space to Virtual Space
4.2.6. Person Physical Volumes in More Detail
4.2.7. Network Capability
4.2.8. Multiple Characters
4.2.9. Interaction with Environment
4.2.10. Level of Detail
4.2.11. Sound at a Point
4.2.12. Sound Shield
4.2.13. Light Shield

# 4.2.1. Introduction

In this chapter, multi-user and networked holodecks are introduced. I first introduce the architecture for multi-user holodecks and then how a networked holodeck could work, building on the theory already introduced. Then I show how the scene is built in virtual main node space and the scene is as per person in the holodeck relative to their position within the main scene and their person physical volume/tile is relative to that position within the scene.

Once these foundations are in place, then I revisit the Person physical volumes in more detail, addressing the network capability with multiple people in the scene and multiple computer characters. Then I show how interaction with the scene could be done and show how, from a Ppv, the user or users interact with the environment with the level of detail being mapped to Ppv circumference space via a Ppv sphere.

Then, building on this, I show how sound could function with Ppvs via the use of sound garbage collection and sound shields. Then lastly I introduce how to ensure a scene is not contaminated with outside light via a light shield.

## 4.2.2. Architecture Overview

From Section 4.1.0, where the Personal physical volume is introduced, which is a cell of volumetric space with 4 m × 4 m × 4 m volume, Fig. 4.2.2.1 shows the breakdown of the Ppv and also shows the Ppv sphere. With this fundamental block in place, it could be determined that in order to make the holodeck commercially viable, a good number of users would have to be possible. In this section, I choose to speculate the design of a volumetric space for 1000 users simultaneously, where each user is able to roam the virtual world at their own will. I have also chosen to set the maximum size of the virtual scene to 100 km × 100 km × 100 km. I have also chosen to set the resolution of the voxel to 20 μm voxel length. This will allow scientific simulations at high resolution and allow maximum usage for the holodeck.

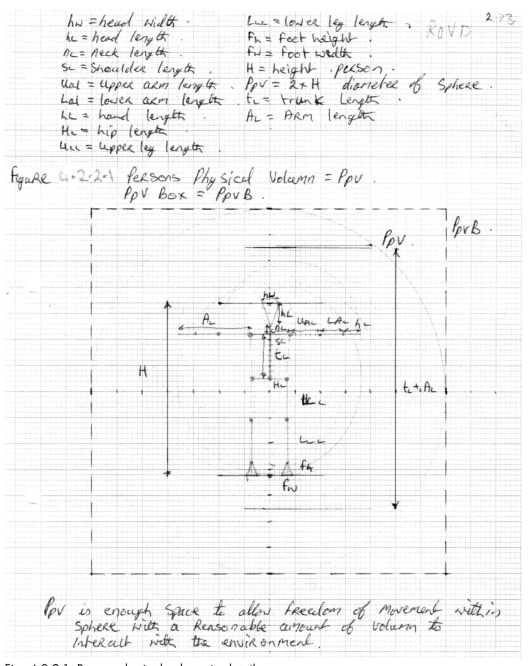

Fig. 4.2.2.1. Person physical volume in detail

**Michael Cloran**

I am going to introduce the multi-user holodeck from the same complex first and then show the networked environment for 1000 users, as from the main node's perspective, similar processing capability and storage has to be achieved for a scene with 100 km × 100 km × 100 km. The size of the main node's scene is big enough to allow 1000 users to have a lot of fun within the same scene.

Fig. 4.2.2.2 shows a floor plan of a volumetric complex with 100 users per layer, with a total of 1000 users. For this volumetric display complex to be possible with 20 µm voxel length,

1m/20 µm = 50,000 voxels per meter

$58m \times 50,000 = 29 \times 10^5$

$29 \times 10^5 \times 29 \times 10^5 \times 29 \times 10^5 = 2.4389 \times 10^{19}$ voxels in the volumetric display complex .

It is worth noting that each voxel has layers of force points and are turned on by setting a bit in the force point buffer where each voxel has such a buffer.

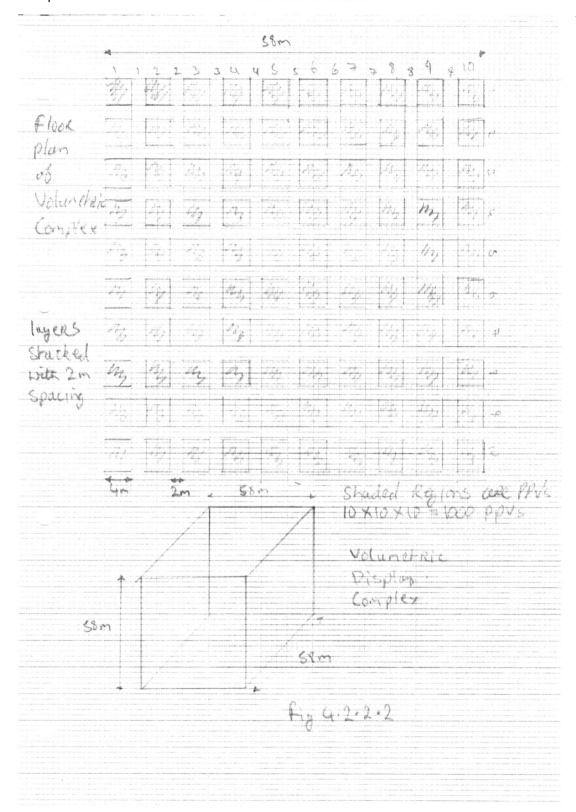

Fig. 4.2.2.2. Volumetric display complex for 1000 users

**Michael Cloran**

Networked multi-user holodeck architecture

Fig. 4.2.2.3 shows a scene with five users in it on the main simulation node and via a network connection, the five people are able to interact with the virtual scene via their personal holodeck Ppv in their own home. I have limited the possible user count to 1000; thus, the actual processing power and storage, as well as the network buffering, would have to be designed specifically for this number.

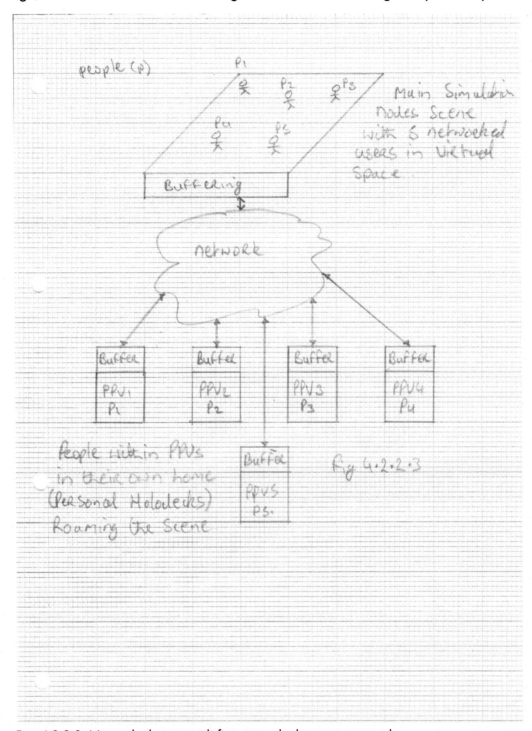

Fig. 4.2.2.3. Networked scene with five networked users as example

From Section 4.1.1, the mappings from voxel space to motherboard thread and core space is still valid; and by looking at the voxel count, the magnitude of the cores can be envisioned.

## 4.2.3. Scene on Main Node

Fig. 4.2.3.1 shows two people within a scene approximately 100 km apart in virtual space. For this to be possible, the main node has to have storage capability for 100 km × 100 km × 100 km of volume with 1 m / 20 μm = 50,000 voxels per meter. That is 50,000 × 100 km = 5,000,000,000 voxels per 100 km length or 5,000,000,000 × 5,000,000,000 = 25,000,000,000,000,000,000 voxels per light plane, which is $25 \times 10^{18} \times 5 \times 10^9 = 1.25 \times 10^{29}$ voxels per volume. But you have to also take into consideration the number of force points per voxel, of which there is a bit buffer for. Each voxel has the capability to be updated at an individual frequency and the possibility of a colour and light level and the possibility of setting the force points within the voxel on rasterisation to a particular newton touch force.

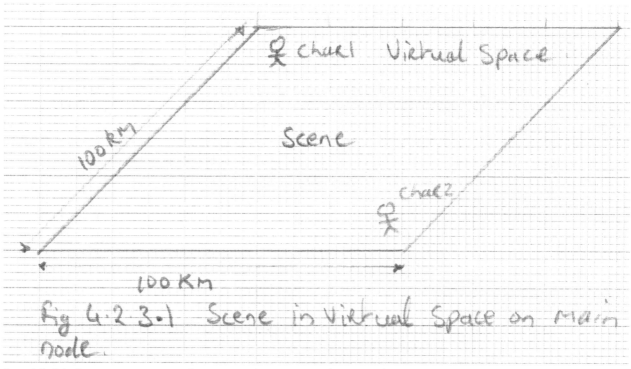

Fig. 4.2.3.1. Basic scene in virtual space on the main simulation node

## 4.2.4. Multiple Ppv Nodes

Fig. 4.2.3.1 shows a basic scene with two people within the scene in virtual space. In reality, these two people are in separate Ppvs within a volumetric display partitioned by a walkway. These people never come together in reality, but in virtual space, one person can enter another person's Ppv. (The people are recreated in virtual space.)

**Michael Cloran**

Fig. 4.2.4.1 shows two people, Character 1 and Character 2, within their own Ppvs. The Ppvs are separated by a space, which is the walkway out of the holodeck if a person decides to exit the holodeck environment.

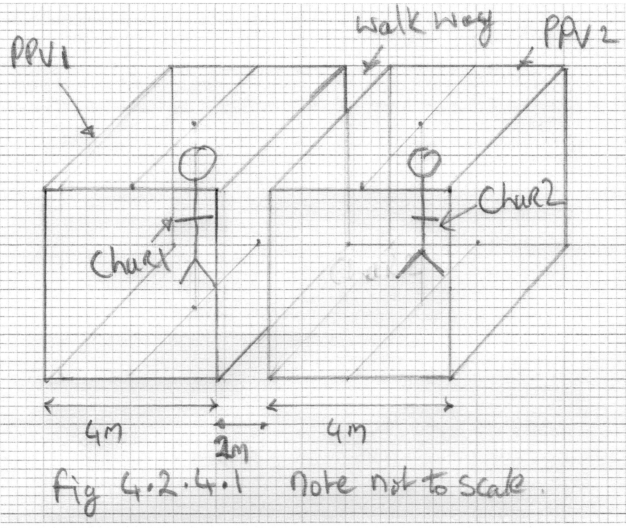

Fig. 4.2.4.1. Volumetric display with two people in it

## 4.2.5. Ppv Physical Space to Virtual Space

For a multi-user holodeck, each person in the holodeck has to be able to roam in any direction within the virtual scene. Thus if the virtual scene is 100 km × 100 km × 100 km (Fig. 4.2.3.1), it can be quickly be seen that space is an issue, and this is dealt with by allowing each person a person physical volume where they interact with the scene rendered in their own personal physical space, which is a reconstruction of the space within the virtual space only with level of detail rendered to the Ppv sphere. This is similar to the scrolling holodeck, only it allows multi-user interaction where each person stays within their own Ppv and they are virtually recreated on the main simulation node. Thus one user can interact with a virtual other user via one user entering the Ppv space of the other user virtually.

It is important to note here that the actual physical space available is 10 m × 4 m × 4 m which is the volumetric space including the exit sign and the walkway. Fig. 4.2.5.1 clarifies this scenario. It is also worth noting that in the scene Fig. 4.2.3.1, in virtual space the two characters are approximately 100 km apart; but in physical space, they are really two meters apart. If they walk towards each other, they will eventually be able to talk to one another and interact with each other (after walking the 100 km). I will deal with sound later.

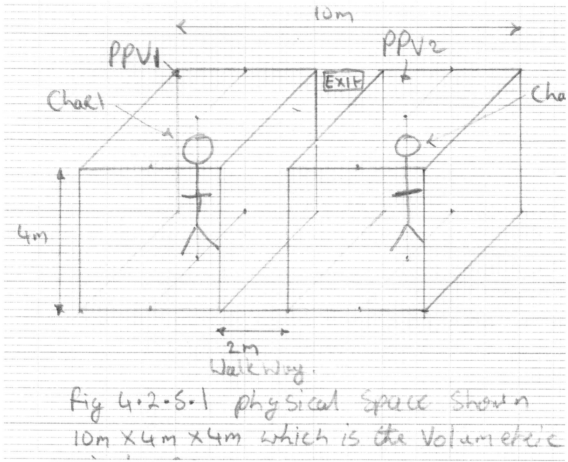

Fig. 4.2.5.1. Volumetric display with two Ppvs and a walkway in the volumetric physical space

## 4.2.6. Personal Physical Volumes in More Detail

From previous chapters, it is obvious that there are certain flaws in the theory of a Ppv in regards to sound and lighting, which I am going to clear up here. For example, if one person starts shouting or even talking, the other person should hear them because in physical space, they are just two meters apart. So to deal with sound, sound detectors will be needed in the Ppv space to measure the amplitude of the sound and the source of the sound. At the circumference of the Ppv sphere, the sound is allowed to exit the scene, only to be garbage collected via being bent towards the floor to be absorbed. With the sound being bent to the floor to stop contamination of sound to this Ppv or to another, a sound shield is used (which blocks sound). The sound, once detected, is ray-traced on the main node; and if, say, the characters are 100 meters apart in virtual space, then via a spatial sound modulator, the sound is recreated in the other person's Ppv at the correct amplitude as if they were 100 meters apart, not 2 meters.

Now that sound is dealt with the Ppv as it stands, the lighting from one Ppv can interfere with the other Ppv and vice versa. This situation would cause severe issues, so to deal with light contamination within a Ppv scene, a light shield is used to reflect light at the bounding box of the Ppv. This can be achieved by just simulating a mirroring effect on the surface of the bounding box. So if a person is on the walkway, all they see is a mirror; but inside the Ppv, the virtual scene on the main node for that tile is recreated with simulated light levels.

## 4.2.7. Network Capability

For a networked holodeck, the physical space required would be just 1 Ppv, and this will allow a user to roam the networked main simulation node's scene (Fig. 4.2.2.3). With a networked holodeck, many users can be in the simulation via their own physical Ppv space in their own home, and they can interact with one another via the virtual scene in virtual space, which is localised to them via their own Ppv space. Also note that if safety is disabled, a person in virtual space can really kill a person within their own Ppv space. It is also possible for a computer-generated character to really kill a person within the holodeck with safety disabled. I will discuss safety later in the holodeck concepts section.

## 4.2.8. Multiple Characters

In a holodeck scene, there can be many people within the simulation as well as several computer-generated characters (Fig. 4.2.8.1). These computer-generated characters have physical bounds where the bounds are the actual characters' physical appearances, which are solid (based on simulation set up by the simulation designer). It is worth noting that in this virtual environment, a bullet from the computer-generated character can kill a real person if safety is disabled, as the bullet is a solid object (based on simulation set up by the simulation designer).

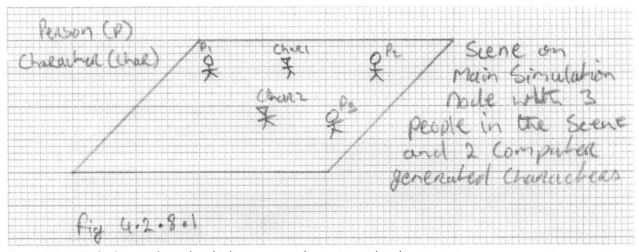

Fig. 4.2.8.1. Multiple people and multiple computer characters within the scene

## 4.2.9. Interaction with Environment

The ability to interact with the environment is limitless based on the imagination of the designer of a simulation. For multi-user holodecks, the concepts of touch and Ppv space is the same only for multi-users within a complex—all users are within a physical volumetric space within a holodeck complex—but for networked holodeck environments, the users may be in their own homes with a volumetric space of 1 Ppv.

## 4.2.10. Level of Detail

Level of detail is used to give a scene depth in that the virtual scene may be 100 km × 100 km × 100 km, but the Ppv physical space is just 4 m × 4 m × 4 m. So if an object is outside the Ppv sphere, it is rendered to it at a level of detail based on the virtual distance from the person in the Ppv to the object. It is worth noting here that inside the Ppv sphere, a local scene is made physical with lighting, touchable objects, etc.

## 4.2.11. Sound at a Point

Sound at a point is spatially modulated sound via the use of a spatial sound modulator. A spatial sound modulator recreates the sound at a point in 3D space such that it has the correct amplitude and basically acts as a virtual speaker at the point in 3D space (Fig. 4.2.11.1).

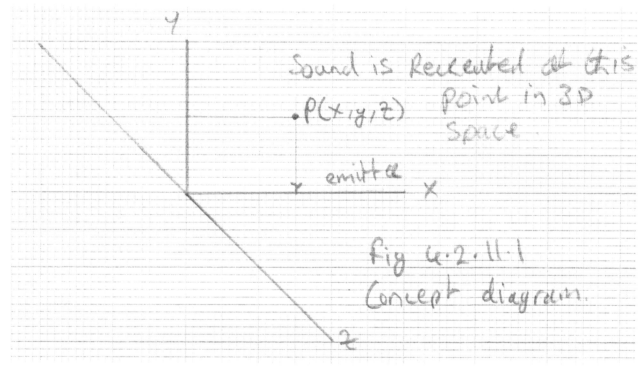

Fig. 4.2.11.1. Sound at a point in 3D space concept diagram

## 4.2.12. Sound Shield

A sound shield is like a force field, but it just stops sound waves. This is used at the Ppv sphere, where the sound can exit only. It cannot re-enter. The Ppv sphere has two sound shields: one on the Ppv sphere and one a short distance away. The first sound shield, which is on the Ppv sphere (Fig. 4.2.12.1), is to stop the sound from re-entering the Ppv space. The second sound shield is to ensure no sound enters or exits the Ppv sphere. The space between the two sound shields is used for bending the sound wave to ground for absorption.

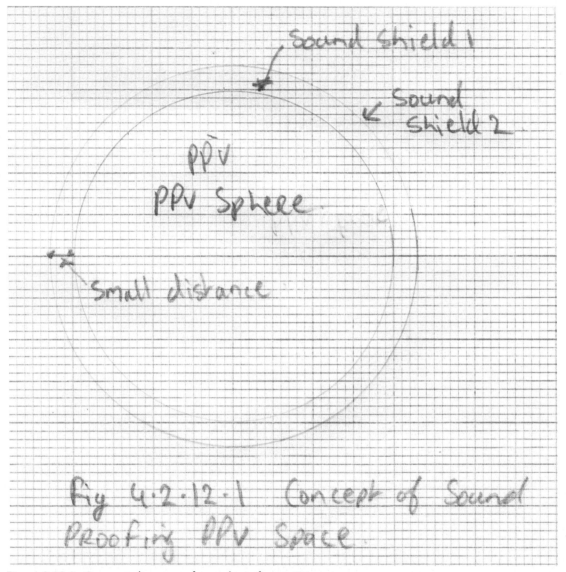

Fig. 4.2.12.1. Concept diagram of soundproofing a Ppv

## 4.2.13. Light Shield

A light shield is just to reflect all light. It can be achieved by simulating a mirror, and thus if a person walks outside the Ppv box, they will see a reflection of themselves in the virtual mirror.

# Holodeck Concepts

4.3.1. Basic Laws of Physics

4.3.2. Basic Wind

4.3.3. Basic Rain

4.3.4. Basic Snow

4.3.5. Character Modelling

4.3.5.1. Clothes for Static and Dynamic Scenes

4.3.5.2. Character Appearance for Static and Dynamic Scenes

4.3.6. Rigging a Character in Volumetric Space

4.3.7. Animation

4.3.8. Computer Characters

4.3.9. Real Characters

4.3.10. Weight

4.3.11. Safety Protocols and No Safety Enabled

# 4.3.1. Basic Laws of Physics

Before I discuss the laws of physics within a holodeck, it is important to brush over some of the major laws of physics in order to get a good view of what is to be implemented for realism at a physical level.

Newton's Laws of Motion

1. *Law 1.* Every body continues in its state of rest or of uniform motion in a straight line, except insofar as it is compelled by impressed forces to change that state.
2. *Law 2.* In an inertial reference frame, the vector sum of forces F on an object is equal to the mass m of that object times the acceleration a of the object: $F = ma$.
3. *Law 3.* To every action, there is an equal and opposite reaction.

Conservation of Mass and Energy

4. *Principle of relativity.* The laws of physics are the same for all inertial reference frames.
5. *Principle of constancy of speed of light.* Light always propagates through a vacuum at a definite velocity, which is independent of the state of motion of the emitting body.

Laws of Thermodynamics

Heat simulations are out of the scope of this book.

Electrostatic Laws

Simulation of electrostatic laws are out of the scope of this book.

From above, the first four laws are relevant to the holodeck scenario for physical simulations. It is important to realise that in implementing these laws, the outcome is just a simulation. I'll explain. All things within a holodeck are virtual but seem real to the person(s) within the simulation environment. Thus in relation to 1 above, if a person has a virtual gun and shoots a virtual projectile, its speed and motion are based on the physics for that simulation event. So to change, for instance, the path of the projectile via an event, the simulation would have to detect the event; then based on the physics simulation engine, it would determine the outcome of the changed path event in order to change the path of the projectile. It is important to note that the outcome of simulation events in the holodeck environment are real (virtually) in that the simulated projectile can kill a person if safety is not enabled.

From 2 above, mass and acceleration in the holodeck are simulated, and the force of an object is based on simulation arguments via simulation events which set the force points within the voxels of the virtual object to give the force.

From 3 above, Newton's third law of motion, an example would be where the terrain in the holodeck is set up such that it has ample force to allow a person to walk upon it. The terrain has an upward force of a certain amount of newtons, and the person walking upon it has a downward force based on weight and force due to gravity. If the upward force of the terrain is the same or higher than the downward force of the person, then the person can safely walk on the terrain.

From 4 above, regarding the principle of relativity, the inertial frame of reference in the holodeck environment is the physical space. The virtual environment virtually moves around the person(s) within the holodeck, and based on simulation events, it sets up artificial forces in order to simulate realism. In order to simulate the principle of relativity, a virtual inertial frame of reference has to be set up in the simulation; and based on outcomes of simulation events, this frame of reference is made to be real to the person(s) within the holodeck.

## 4.3.2. Basic Airflow/Wind

To simulate normal wind flow, every voxel to be used within the flow would have to give a certain force in a direction to give the experience of wind where the newton force of the force points within the voxels are set to a value which is appropriate for air.

Example Simulation

Imagine getting into a plane, being flown to 30,000 feet, and then jumping out with a parachute and free falling with increasing velocity towards the ground until terminal velocity is achieved. You then continue falling until the parachute is activated, at which the air resistance increases, severely decreasing the velocity of your descent to the ground.

This scenario can be simulated in a holodeck. In particular, a networked holodeck would suit the purpose, where roaming the virtual scene is possible. To begin the simulation, you would walk up the steps into the plane to be handed a virtual parachute to put on. Then, after a time to be flown to 30,000 feet, the door would open and you could jump out.

It is important to realise that the physical space available is a Ppvs space which is 4 m × 4 m × 4 m, so when you jump out, you are held in free space and the world moves around you, with the air resistance simulated as an airflow around you. Thus the sensation of the flow of air would be possible.

When the parachute is pulled, the speed of the flow of air decreases, and you glide towards the ground. The parachute is virtual, and when you pull, the parachute bounds may be outside the bounds of the Ppv sphere. Thus, part of it would be textured on to the Ppv sphere.

Wind flow would also be interesting in the case of a person climbing a virtual cliff where as the altitude increases, the wind could get more severe or even with gusts.

## 4.3.3. Basic Rain

Rain is tiny droplets falling from the sky. On hitting something, the small droplets give a slight force on the object it hits. The droplets break up into smaller droplets on the object and into a small flow in some cases, which wets the surface of the hit object.

Rain in a simulation is not realistic if all the rain fell straight down in the same direction without wind. So to simulate rain with wind, there would have to be a wind flow direction; and with this, the droplets would fall according to the force exerted on the droplet to make the droplets flow with the wind.

To simulate a rain droplet, a tiny sphere with the circumference with the material of water can be set to a touch force of water. On hitting/colliding with an object, the droplet breaks up; and depending on the simulation, it breaks into smaller droplets or creates a flow on the surface of the object it hits, where the flow of the water would have the touch force of water.

## 4.3.4. Basic Snow

Snow is water vapor frozen into ice crystals.

To simulate it, the ice crystal has a material structure. By setting up a virtual molecular structure at the surface, reflection takes place at the colour of the snowflake, which is white, and by setting the force points of the virtual crystal to a touch force of snow, you could get the feel of touching snow.

As before with rain, snow would be not realistic if it just fell straight down in a simulation, so to combine it with wind would make it more realistic.

## 4.3.5. Character Modelling

Character modelling in a holodeck is true 3D editing in volumetric space. A character could be formed from a template cube of the material type of a soft material. Then with this template, you could add to the model or remove from the model material with various tools. It would be possible to change the colour and texture of the material. Then a finished model template could use this as a skin.

This would work in the holodeck via the sculptor setting the cube of size. Then the material would be chosen to be easy to deform and to add to for ease of sculpting. Then when the sculpting is finished, the material type and colour can be changed, and this could be used as template to be passed to the animator.

## 4.3.5.1. Clothes for Static and Dynamic Scenes

In the types of holodeck described in this book, a requirement in clothes to wear would be that they would be tight-fitting so the simulator can easily project clothing on to your person for the various simulations.

## 4.3.5.2. Character Appearance for Static and Dynamic Scenes

For a static scene, the physical space for the scene is determined by the physical space available; and thus if the person in the scene is an orc, then the particular physical characteristics of the orc is projected and blended on to the character in the scene so they seem to be an orc. This is true also for dynamic scenes where if you look at yourself while in the Ppv, you would seem to be an orc; and on the main simulation node, your virtual self would be an orc.

## 4.3.6. Rigging a Character in Volumetric Space

Once the template character has been modelled, in order to bring the static model to life, two approaches are possible for a holodeck.

1.  Modelling the character based on bone and muscle movement
2.  Using the static rig and setting up a deformation rig the same way it's done in modern 3D software for a 2D screen

1.  Modelling a character based on bone and muscle movement
With the template rig, you have a guide of how the finished model should look like. From this model, you could build from the inside out, adding bone and muscle. When the muscle is lifting weight, it contracts. A deformation of the muscle is seen when the muscle lets go of the weight, and an eccentric contraction is seen where the muscle relaxes.

2.  Rigging a model with current 3D software
This would require setting up deformation points and building a deformation rig around the static model to bring it to life. Depending on the requirement for the simulation, a deformation based on a frame could be 60 frames per second, or higher frame rates could be used.

## 4.3.7. Animation

Animation of a character could be done at 60 frames per second, where each frame would increment in a deformation by a predefined amount, which is based on the frame number. In each frame, the character is moved a certain amount; and when the frames are flipped, movement is seen.

## 4.3.8. Computer Characters

Computer or animated characters are based on the character-modelling, rigging, and frame-sequencing pipeline.

## 4.3.9. Real Characters

Real characters or people can have their natural appearance or a projected blending look and feel. From the example above, just imagine a person with an orc appearance projected on to him or her.

## 4.3.10. Weight

An object in virtual space does not have a physical mass. Thus the mass is calculated, and a simulated reality (it does not have to hold true). I'll explain. An object in virtual space is a set of spatially positioned moments that are set up in order to simulate the material in question. The existence of weight can be simulated in that an object which is moveable can exert resistance (simulated) to the body trying to move the object. This is done via adjusting the force of resistance by altering the magnitude of the spatial force points. Weight can be measured in newtons of resistance to a body and can be set up for particular simulations. If the person is not able to lift the weight, then an animation of the weight falling to the floor is seen. If they can lift the object, their muscles constantly feel the downward force of the object's weight. This is a simulation of force downwards to the body lifting the weight.

## 4.3.11. Safety Protocols and No Safety Enabled

It is important to realise that in a holodeck, the environment simulates the real world. Therefore, there are many ways in which a person can be injured or killed while inside this environment.

I am going to break the type of dangers into three types:

1. falling
2. weights
3. virtual bullets

1. Falling
When falling, you can hurt yourself because in this environment, you may not realise the fall until you have done it. Just imagine falling through the floorboards in a simulation in a game once you walk. This would need safety, as you don't realise the fall until you are falling.

In the simulation of free fall, if you don't activate the parachute, you are falling. The simulation could end safely where you would reach a point where you fall no more and you are put upright.

If in a falling situation where you are twisting and turning within the fall, it may be important to stop rotations before cushioning the fall. In a simulation within a Ppv, the dimensions are 4 m × 4 m × 4 m; so at most, you could fall approximately half of this because it is a requirement to keep a person approximately in the middle of the Ppv. The cushion would have to deal with a 2 m fall at most. This cushion would be a virtual cushion in that the voxels within its volume would be set to the newton force of a cushion to halt the fall.

2. Weights

When a person is lifting a virtual weight, it is possible to do damage. A suggested approach is to have the weight set to a small weight in order to just give the sensation of weight.

3.  Virtual Bullets

Virtual bullets can kill because bullets move at speed and have, on its surface, an appearance of solidness. Thus it would appear to be real. Stopping the bullet is complex for safety reasons, as the bullet is a simulation and is created in an animation sequence. Normal armour won't stop the computer from animating the bullets moving from A to B. So to stop a bullet from damaging a person within the path A to B, it would have to be possible for a personal shield to form around the person with the sole purpose of stopping the fields generated by emitters that project the bullet from being updated for the volume of the person within the personal shield. Thus, within this volume, the bullet would not exist, but it could appear to exit the volume at the far side, as the simulation is set for path A to B.

# Conclusion

This book covers a lot of ground from optical processor functions to instruction sets, to volumetrics, and then to the evolution of volumetric theory, to holodeck theory. In covering this material, I introduce a test case holodeck which is a 100 km × 100 km × 100 km data centre virtual scene which can be interfaced with through personal holodecks which are 4 m × 4 m × 4 m in size.

I introduce a way for holodecks to function in a multi-user way—Ppv theory, which entails sound at a point of which the level of sound heard is determined by the virtual distance from the sound source. I also cover ground on shield theories—sound shields, light shields, and for safety, a personal shield. Texturing and levels of detail are also touched upon, along with a new way of creating a voxel of information for a volumetric display. Solidness to touch is also covered.

Choosing the force point count to be 10 per axis within a $20 \times 10^{-6}$ size voxel might not give the desired effect, but it does give an entry-level look at such a system. So my approach to this is to start with 10 per axis, then increase this to $1 \times 10^3$, and then increase this to $1 \times 10^6$, then to $1 \times 10^9$. I understand that the later values per axis might not be feasible to fabricate. The implication of this is that a voxel would have between $1 \times 10^9$ force points per voxel volume and $1 \times 10^{27}$ force points per voxel volume. This would increase the threads and cores—and ultimately, the RAM for the system—to colossal amounts and would bring the already massive system to a new level of scale.

This project was a long journey, and on the path, I have had to learn about a lot of different technologies. In doing so, I believe that once the theories within this book are proofed formally, it will lead to holodeck design. And at that time, the technologies within this book will be globally used on a daily basis.

# Appendix A

The purpose of this appendix is to introduce the reader to the component images used in the simulator to aid in the writing of this book.

Fig. A.1. Two-input AND gate which performs the logical AND operation

Truth Table A.1

| A | B | Output |
|---|---|--------|
| $\lambda_1[0]$ | $\lambda_2[0]$ | $\lambda_1[0]$ |
| $\lambda_1[1]$ | $\lambda_2[0]$ | $\lambda_1[0]$ |
| $\lambda_1[0]$ | $\lambda_2[1]$ | $\lambda_1[0]$ |
| $\lambda_1[1]$ | $\lambda_2[1]$ | $\lambda_1[1]$ |

Note that in the truth tables, it is assumed that the wavelength at pin 1 is the wavelength of the output. This helps keep the wavelengths aligned.

In the simulator, I am using an ideal simulation model where the input wavelength is an integer which symbolises the characteristics of the wavelength and the number within the brackets symbolise that the wavelength is intensity modulated with this value.

Fig. A.2. Three-input AND gate which performs the logical AND operation

Truth Table A.2

| A | B | C | Output |
|---|---|---|--------|
| $\lambda_1[0]$ | $\lambda_2[0]$ | $\lambda_3[0]$ | $\lambda_1[0]$ |
| $\lambda_1[1]$ | $\lambda_2[0]$ | $\lambda_3[0]$ | $\lambda_1[0]$ |
| $\lambda_1[0]$ | $\lambda_2[1]$ | $\lambda_3[0]$ | $\lambda_1[0]$ |

| A | B | C | Output |
|---|---|---|---|
| $\lambda_1[1]$ | $\lambda_2[1]$ | $\lambda_3[0]$ | $\lambda_1[0]$ |
| $\lambda_1[0]$ | $\lambda_2[0]$ | $\lambda_3[1]$ | $\lambda_1[0]$ |
| $\lambda_1[1]$ | $\lambda_2[0]$ | $\lambda_3[1]$ | $\lambda_1[0]$ |
| $\lambda_1[0]$ | $\lambda_2[1]$ | $\lambda_3[1]$ | $\lambda_1[0]$ |
| $\lambda_1[1]$ | $\lambda_2[1]$ | $\lambda_3[1]$ | $\lambda_1[1]$ |

Fig. A.3. Four-input AND gate which performs the logical AND operation

Truth Table A.3

| A | B | C | D | Output |
|---|---|---|---|---|
| $\lambda_1[0]$ | $\lambda_2[0]$ | $\lambda_3[0]$ | $\lambda_4[0]$ | $\lambda_1[0]$ |
| $\lambda_1[1]$ | $\lambda_2[0]$ | $\lambda_3[0]$ | $\lambda_4[0]$ | $\lambda_1[0]$ |
| .. | .. | .. | .. | .. |
| .. | .. | .. | .. | .. |
| $\lambda_1[1]$ | $\lambda_2[1]$ | $\lambda_3[1]$ | $\lambda_4[1]$ | $\lambda_1[1]$ |

Fig. A.4. Five-input AND gate which performs the logical AND operation

Truth Table A.4

| A | B | C | D | E | Output |
|---|---|---|---|---|---|
| $\lambda_1[0]$ | $\lambda_2[0]$ | $\lambda_3[0]$ | $\lambda_4[0]$ | $\lambda_5[0]$ | $\lambda_1[0]$ |
| $\lambda_1[1]$ | $\lambda_2[0]$ | $\lambda_3[0]$ | $\lambda_4[0]$ | $\lambda_5[0]$ | $\lambda_1[0]$ |
| .. | .. | .. | .. | .. | .. |

| A | B | C | D | E | Output |
|---|---|---|---|---|---|
| .. | .. | .. | .. | .. | .. |
| $\lambda_1[1]$ | $\lambda_2[1]$ | $\lambda_3[1]$ | $\lambda_4[1]$ | $\lambda_5[1]$ | $\lambda_1[1]$ |

Fig. A.5. Six-input AND gate which performs the logical AND operation

Truth Table A.5

| A | B | C | D | E | F | Output |
|---|---|---|---|---|---|---|
| $\lambda_1[0]$ | $\lambda_2[0]$ | $\lambda_3[0]$ | $\lambda_4[0]$ | $\lambda_5[0]$ | $\lambda_6[0]$ | $\lambda_1[0]$ |
| $\lambda_1[1]$ | $\lambda_2[0]$ | $\lambda_3[0]$ | $\lambda_4[0]$ | $\lambda_5[0]$ | $\lambda_6[0]$ | $\lambda_1[0]$ |
| .. | .. | .. | .. | .. | .. | .. |
| .. | .. | .. | .. | .. | .. | .. |
| $\lambda_1[1]$ | $\lambda_2[1]$ | $\lambda_3[1]$ | $\lambda_4[1]$ | $\lambda_5[1]$ | $\lambda_6[1]$ | $\lambda_1[1]$ |

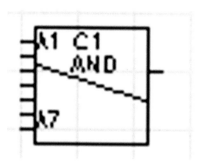

Fig. A.6. Seven-input AND gate which performs the logical AND operation

Truth Table A.6

| A | B | C | D | E | F | G | Output |
|---|---|---|---|---|---|---|---|
| $\lambda_1[0]$ | $\lambda_2[0]$ | $\lambda_3[0]$ | $\lambda_4[0]$ | $\lambda_5[0]$ | $\lambda_6[0]$ | $\lambda_7[0]$ | $\lambda_1[0]$ |
| $\lambda_1[1]$ | $\lambda_2[0]$ | $\lambda_3[0]$ | $\lambda_4[0]$ | $\lambda_5[0]$ | $\lambda_6[0]$ | $\lambda_7[0]$ | $\lambda_1[0]$ |
| .. | .. | .. | .. | .. | .. | .. | .. |
| .. | .. | .. | .. | .. | .. | .. | .. |
| $\lambda_1[1]$ | $\lambda_2[1]$ | $\lambda_3[1]$ | $\lambda_4[1]$ | $\lambda_5[1]$ | $\lambda_6[1]$ | $\lambda_7[1]$ | $\lambda_1[1]$ |

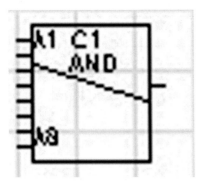

Fig. A.7. Eight-input AND gate which performs the logical AND operation

Truth Table A.7

| A | B | C | D | E | F | G | H | Output |
|---|---|---|---|---|---|---|---|---|
| $\lambda_1[0]$ | $\lambda_2[0]$ | $\lambda_3[0]$ | $\lambda_4[0]$ | $\lambda_5[0]$ | $\lambda_6[0]$ | $\lambda_7[0]$ | $\lambda_8[0]$ | $\lambda_1[0]$ |
| $\lambda_1[1]$ | $\lambda_2[0]$ | $\lambda_3[0]$ | $\lambda_4[0]$ | $\lambda_5[0]$ | $\lambda_6[0]$ | $\lambda_7[0]$ | $\lambda_8[0]$ | $\lambda_1[0]$ |
| .. | .. | .. | .. | .. | .. | .. | .. | .. |
| .. | .. | .. | .. | .. | .. | .. | .. | .. |
| $\lambda_1[1]$ | $\lambda_2[1]$ | $\lambda_3[1]$ | $\lambda_4[1]$ | $\lambda_5[1]$ | $\lambda_6[1]$ | $\lambda_7[1]$ | $\lambda_8[1]$ | $\lambda_1[1]$ |

Fig. A.8. Two-input NAND gate which performs the logical NAND operation

Truth Table A.8

| A | B | Output |
|---|---|---|
| $\lambda_1[0]$ | $\lambda_2[0]$ | $\lambda_1[1]$ |
| $\lambda_1[1]$ | $\lambda_2[0]$ | $\lambda_1[1]$ |
| $\lambda_1[0]$ | $\lambda_2[1]$ | $\lambda_1[1]$ |
| $\lambda_1[1]$ | $\lambda_2[1]$ | $\lambda_1[0]$ |

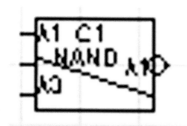

Fig. A.9. Three-input NAND gate which performs the logical NAND operation

Truth Table A.9

| A | B | C | Output |
|---|---|---|---|
| $\lambda_1[0]$ | $\lambda_2[0]$ | $\lambda_3[0]$ | $\lambda_1[1]$ |
| $\lambda_1[1]$ | $\lambda_2[0]$ | $\lambda_3[0]$ | $\lambda_1[1]$ |
| $\lambda_1[0]$ | $\lambda_2[1]$ | $\lambda_3[0]$ | $\lambda_1[1]$ |
| $\lambda_1[1]$ | $\lambda_2[1]$ | $\lambda_3[0]$ | $\lambda_1[1]$ |
| $\lambda_1[0]$ | $\lambda_2[0]$ | $\lambda_3[1]$ | $\lambda_1[1]$ |
| $\lambda_1[1]$ | $\lambda_2[0]$ | $\lambda_3[1]$ | $\lambda_1[1]$ |
| $\lambda_1[0]$ | $\lambda_2[1]$ | $\lambda_3[1]$ | $\lambda_1[1]$ |
| $\lambda_1[1]$ | $\lambda_2[1]$ | $\lambda_3[1]$ | $\lambda_1[0]$ |

Fig. A.10. Four-input NAND gate which performs the logical NAND operation

Truth Table A.10

| A | B | C | D | Output |
|---|---|---|---|---|
| $\lambda_1[0]$ | $\lambda_2[0]$ | $\lambda_3[0]$ | $\lambda_4[0]$ | $\lambda_1[1]$ |
| $\lambda_1[1]$ | $\lambda_2[0]$ | $\lambda_3[0]$ | $\lambda_4[0]$ | $\lambda_1[1]$ |
| .. | .. | .. | .. | .. |
| .. | .. | .. | .. | .. |
| $\lambda_1[1]$ | $\lambda_2[1]$ | $\lambda_3[1]$ | $\lambda_4[1]$ | $\lambda_1[0]$ |

Fig. A.11. Five-input NAND gate which performs the logical NAND operation

Truth Table A.11

| A | B | C | D | E | Output |
|---|---|---|---|---|---|
| $\lambda_1[0]$ | $\lambda_2[0]$ | $\lambda_3[0]$ | $\lambda_4[0]$ | $\lambda_5[0]$ | $\lambda_1[1]$ |
| $\lambda_1[1]$ | $\lambda_2[0]$ | $\lambda_3[0]$ | $\lambda_4[0]$ | $\lambda_5[0]$ | $\lambda_1[1]$ |
| .. | .. | .. | .. | .. | .. |
| .. | .. | .. | .. | .. | .. |
| $\lambda_1[1]$ | $\lambda_2[1]$ | $\lambda_3[1]$ | $\lambda_4[1]$ | $\lambda_5[1]$ | $\lambda_1[0]$ |

Fig. A.12. Six-input NAND gate which performs the logical NAND operation

Truth Table A.12

| A | B | C | D | E | F | Output |
|---|---|---|---|---|---|---|
| $\lambda_1[0]$ | $\lambda_2[0]$ | $\lambda_3[0]$ | $\lambda_4[0]$ | $\lambda_5[0]$ | $\lambda_6[0]$ | $\lambda_1[1]$ |
| $\lambda_1[1]$ | $\lambda_2[0]$ | $\lambda_3[0]$ | $\lambda_4[0]$ | $\lambda_5[0]$ | $\lambda_6[0]$ | $\lambda_1[1]$ |
| .. | .. | .. | .. | .. | .. | .. |
| .. | .. | .. | .. | .. | .. | .. |
| $\lambda_1[1]$ | $\lambda_2[1]$ | $\lambda_3[1]$ | $\lambda_4[1]$ | $\lambda_5[1]$ | $\lambda_6[1]$ | $\lambda_1[0]$ |

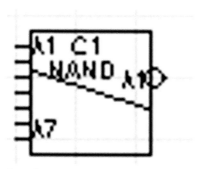

Fig. A.13. Seven-input NAND gate which performs the logical NAND operation

Truth Table A.13

| A | B | C | D | E | F | G | Output |
|---|---|---|---|---|---|---|--------|
| $\lambda_1[0]$ | $\lambda_2[0]$ | $\lambda_3[0]$ | $\lambda_4[0]$ | $\lambda_5[0]$ | $\lambda_6[0]$ | $\lambda_7[0]$ | $\lambda_1[1]$ |
| $\lambda_1[1]$ | $\lambda_2[0]$ | $\lambda_3[0]$ | $\lambda_4[0]$ | $\lambda_5[0]$ | $\lambda_6[0]$ | $\lambda_7[0]$ | $\lambda_1[1]$ |
| .. | .. | .. | .. | .. | .. | .. | .. |
| .. | .. | .. | .. | .. | .. | .. | .. |
| $\lambda_1[1]$ | $\lambda_2[1]$ | $\lambda_3[1]$ | $\lambda_4[1]$ | $\lambda_5[1]$ | $\lambda_6[1]$ | $\lambda_7[1]$ | $\lambda_1[0]$ |

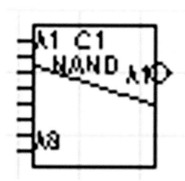

Fig. A.14. Eight-input NAND gate which performs the logical NAND operation

Truth Table A.14

| A | B | C | D | E | F | G | H | Output |
|---|---|---|---|---|---|---|---|--------|
| $\lambda_1[0]$ | $\lambda_2[0]$ | $\lambda_3[0]$ | $\lambda_4[0]$ | $\lambda_5[0]$ | $\lambda_6[0]$ | $\lambda_7[0]$ | $\lambda_8[0]$ | $\lambda_1[1]$ |
| $\lambda_1[1]$ | $\lambda_2[0]$ | $\lambda_3[0]$ | $\lambda_4[0]$ | $\lambda_5[0]$ | $\lambda_6[0]$ | $\lambda_7[0]$ | $\lambda_8[0]$ | $\lambda_1[1]$ |
| .. | .. | .. | .. | .. | .. | .. | .. | .. |
| .. | .. | .. | .. | .. | .. | .. | .. | .. |
| $\lambda_1[1]$ | $\lambda_2[1]$ | $\lambda_3[1]$ | $\lambda_4[1]$ | $\lambda_5[1]$ | $\lambda_6[1]$ | $\lambda_7[1]$ | $\lambda_8[1]$ | $\lambda_1[0]$ |

Fig. A.15. Two-input NOR gate which performs the logical NOR operation

Truth Table A.15

| A | B | Output |
|---|---|---|
| $\lambda_1[0]$ | $\lambda_2[0]$ | $\lambda_1[1]$ |
| $\lambda_1[1]$ | $\lambda_2[0]$ | $\lambda_1[0]$ |
| $\lambda_1[0]$ | $\lambda_2[1]$ | $\lambda_1[0]$ |
| $\lambda_1[1]$ | $\lambda_2[1]$ | $\lambda_1[0]$ |

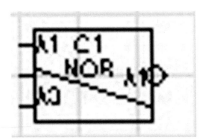

Fig. A.16. Three-input NOR gate which performs the logical NOR operation

Truth Table A.16

| A | B | C | Output |
|---|---|---|---|
| $\lambda_1[0]$ | $\lambda_2[0]$ | $\lambda_3[0]$ | $\lambda_1[1]$ |
| $\lambda_1[1]$ | $\lambda_2[0]$ | $\lambda_3[0]$ | $\lambda_1[0]$ |
| $\lambda_1[0]$ | $\lambda_2[1]$ | $\lambda_3[0]$ | $\lambda_1[0]$ |
| $\lambda_1[1]$ | $\lambda_2[1]$ | $\lambda_3[0]$ | $\lambda_1[0]$ |
| $\lambda_1[0]$ | $\lambda_2[0]$ | $\lambda_3[1]$ | $\lambda_1[0]$ |
| $\lambda_1[1]$ | $\lambda_2[0]$ | $\lambda_3[1]$ | $\lambda_1[0]$ |
| $\lambda_1[0]$ | $\lambda_2[1]$ | $\lambda_3[1]$ | $\lambda_1[0]$ |
| $\lambda_1[1]$ | $\lambda_2[1]$ | $\lambda_3[1]$ | $\lambda_1[0]$ |

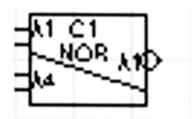

Fig. A.17. Four-input NOR gate which performs the logical NOR operation

Truth Table A.17

| A | B | C | D | Output |
|---|---|---|---|---|
| $\lambda_1[0]$ | $\lambda_2[0]$ | $\lambda_3[0]$ | $\lambda_4[0]$ | $\lambda_1[1]$ |
| $\lambda_1[1]$ | $\lambda_2[0]$ | $\lambda_3[0]$ | $\lambda_4[0]$ | $\lambda_1[0]$ |
| .. | .. | .. | .. | .. |
| .. | .. | .. | .. | .. |
| $\lambda_1[1]$ | $\lambda_2[1]$ | $\lambda_3[1]$ | $\lambda_4[1]$ | $\lambda_1[0]$ |

Fig. A.18. Five-input NOR gate which performs the logical NOR operation

Truth Table A.18

| A | B | C | D | E | Output |
|---|---|---|---|---|---|
| $\lambda_1[0]$ | $\lambda_2[0]$ | $\lambda_3[0]$ | $\lambda_4[0]$ | $\lambda_5[0]$ | $\lambda_1[1]$ |
| $\lambda_1[1]$ | $\lambda_2[0]$ | $\lambda_3[0]$ | $\lambda_4[0]$ | $\lambda_5[0]$ | $\lambda_1[0]$ |
| .. | .. | .. | .. | .. | .. |
| .. | .. | .. | .. | .. | .. |
| $\lambda_1[1]$ | $\lambda_2[1]$ | $\lambda_3[1]$ | $\lambda_4[1]$ | $\lambda_5[1]$ | $\lambda_1[0]$ |

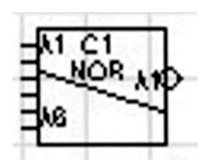

Fig. A.19. Six-input NOR gate which performs the logical NOR operation

Truth Table A.19

| A | B | C | D | E | F | Output |
|---|---|---|---|---|---|--------|
| $\lambda_1[0]$ | $\lambda_2[0]$ | $\lambda_3[0]$ | $\lambda_4[0]$ | $\lambda_5[0]$ | $\lambda_6[0]$ | $\lambda_1[1]$ |
| $\lambda_1[1]$ | $\lambda_2[0]$ | $\lambda_3[0]$ | $\lambda_4[0]$ | $\lambda_5[0]$ | $\lambda_6[0]$ | $\lambda_1[0]$ |
| .. | .. | .. | .. | .. | .. | .. |
| .. | .. | .. | .. | .. | .. | .. |
| $\lambda_1[1]$ | $\lambda_2[1]$ | $\lambda_3[1]$ | $\lambda_4[1]$ | $\lambda_5[1]$ | $\lambda_6[1]$ | $\lambda_1[0]$ |

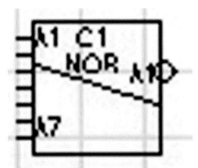

Fig. A.20. Seven-input NOR gate which performs the logical NOR operation

Truth Table A.20

| A | B | C | D | E | F | G | Output |
|---|---|---|---|---|---|---|--------|
| $\lambda_1[0]$ | $\lambda_2[0]$ | $\lambda_3[0]$ | $\lambda_4[0]$ | $\lambda_5[0]$ | $\lambda_6[0]$ | $\lambda_7[0]$ | $\lambda_1[1]$ |
| $\lambda_1[1]$ | $\lambda_2[0]$ | $\lambda_3[0]$ | $\lambda_4[0]$ | $\lambda_5[0]$ | $\lambda_6[0]$ | $\lambda_7[0]$ | $\lambda_1[0]$ |
| .. | .. | .. | .. | .. | .. | .. | .. |
| .. | .. | .. | .. | .. | .. | .. | .. |
| $\lambda_1[1]$ | $\lambda_2[1]$ | $\lambda_3[1]$ | $\lambda_4[1]$ | $\lambda_5[1]$ | $\lambda_6[1]$ | $\lambda_7[1]$ | $\lambda_1[0]$ |

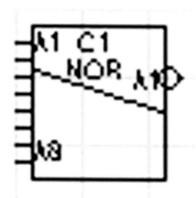

Fig. A.21. Eight-input NOR gate which performs the logical NOR operation

Truth Table A.21

| A | B | C | D | E | F | G | H | Output |
|---|---|---|---|---|---|---|---|--------|
| $\lambda_1[0]$ | $\lambda_2[0]$ | $\lambda_3[0]$ | $\lambda_4[0]$ | $\lambda_5[0]$ | $\lambda_6[0]$ | $\lambda_7[0]$ | $\lambda_8[0]$ | $\lambda_1[1]$ |
| $\lambda_1[1]$ | $\lambda_2[0]$ | $\lambda_3[0]$ | $\lambda_4[0]$ | $\lambda_5[0]$ | $\lambda_6[0]$ | $\lambda_7[0]$ | $\lambda_8[0]$ | $\lambda_1[0]$ |
| .. | .. | .. | .. | .. | .. | .. | .. | .. |
| .. | .. | .. | .. | .. | .. | .. | .. | .. |
| $\lambda_1[1]$ | $\lambda_2[1]$ | $\lambda_3[1]$ | $\lambda_4[1]$ | $\lambda_5[1]$ | $\lambda_6[1]$ | $\lambda_7[1]$ | $\lambda_8[1]$ | $\lambda_1[0]$ |

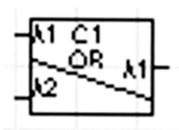

Fig. A.22. Two-input OR gate which performs the logical OR operation

Truth Table A.22

| A | B | Output |
|---|---|--------|
| $\lambda_1[0]$ | $\lambda_2[0]$ | $\lambda_1[0]$ |
| $\lambda_1[1]$ | $\lambda_2[0]$ | $\lambda_1[1]$ |
| $\lambda_1[0]$ | $\lambda_2[1]$ | $\lambda_1[1]$ |
| $\lambda_1[1]$ | $\lambda_2[1]$ | $\lambda_1[1]$ |

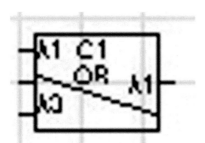

Fig. A.23. Three-input OR gate which performs the logical OR operation

Truth Table A.23

| A | B | C | Output |
|---|---|---|---|
| $\lambda_1[0]$ | $\lambda_2[0]$ | $\lambda_3[0]$ | $\lambda_1[0]$ |
| $\lambda_1[1]$ | $\lambda_2[0]$ | $\lambda_3[0]$ | $\lambda_1[1]$ |
| $\lambda_1[0]$ | $\lambda_2[1]$ | $\lambda_3[0]$ | $\lambda_1[1]$ |
| $\lambda_1[1]$ | $\lambda_2[1]$ | $\lambda_3[0]$ | $\lambda_1[1]$ |
| $\lambda_1[0]$ | $\lambda_2[0]$ | $\lambda_3[1]$ | $\lambda_1[1]$ |
| $\lambda_1[1]$ | $\lambda_2[0]$ | $\lambda_3[1]$ | $\lambda_1[1]$ |
| $\lambda_1[0]$ | $\lambda_2[1]$ | $\lambda_3[1]$ | $\lambda_1[1]$ |
| $\lambda_1[1]$ | $\lambda_2[1]$ | $\lambda_3[1]$ | $\lambda_1[1]$ |

Fig. A.24. Four-input OR gate which performs the logical OR operation

Truth Table A.24

| A | B | C | D | Output |
|---|---|---|---|---|
| $\lambda_1[0]$ | $\lambda_2[0]$ | $\lambda_3[0]$ | $\lambda_4[0]$ | $\lambda_1[0]$ |
| $\lambda_1[1]$ | $\lambda_2[0]$ | $\lambda_3[0]$ | $\lambda_4[0]$ | $\lambda_1[1]$ |
| .. | .. | .. | .. | $\lambda_1[1]$ |
| .. | .. | .. | .. | $\lambda_1[1]$ |
| $\lambda_1[1]$ | $\lambda_2[1]$ | $\lambda_3[1]$ | $\lambda_4[1]$ | $\lambda_1[1]$ |

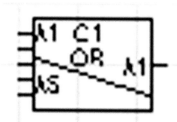

Fig. A.25. Five-input OR gate which performs the logical OR operation

Truth Table A.25

| A | B | C | D | E | Output |
|---|---|---|---|---|---|
| $\lambda_1[0]$ | $\lambda_2[0]$ | $\lambda_3[0]$ | $\lambda_4[0]$ | $\lambda_5[0]$ | $\lambda_1[0]$ |
| $\lambda_1[1]$ | $\lambda_2[0]$ | $\lambda_3[0]$ | $\lambda_4[0]$ | $\lambda_5[0]$ | $\lambda_1[1]$ |
| .. | .. | .. | .. | .. | $\lambda_1[1]$ |
| .. | .. | .. | .. | .. | $\lambda_1[1]$ |
| $\lambda_1[1]$ | $\lambda_2[1]$ | $\lambda_3[1]$ | $\lambda_4[1]$ | $\lambda_5[1]$ | $\lambda_1[1]$ |

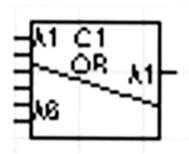

Fig. A.26. Six-input OR gate which performs the logical OR operation

Truth Table A.26

| A | B | C | D | E | F | Output |
|---|---|---|---|---|---|---|
| $\lambda_1[0]$ | $\lambda_2[0]$ | $\lambda_3[0]$ | $\lambda_4[0]$ | $\lambda_5[0]$ | $\lambda_6[0]$ | $\lambda_1[0]$ |
| $\lambda_1[1]$ | $\lambda_2[0]$ | $\lambda_3[0]$ | $\lambda_4[0]$ | $\lambda_5[0]$ | $\lambda_6[0]$ | $\lambda_1[1]$ |
| .. | .. | .. | .. | .. | .. | $\lambda_1[1]$ |
| .. | .. | .. | .. | .. | .. | $\lambda_1[1]$ |
| $\lambda_1[1]$ | $\lambda_2[1]$ | $\lambda_3[1]$ | $\lambda_4[1]$ | $\lambda_5[1]$ | $\lambda_6[1]$ | $\lambda_1[1]$ |

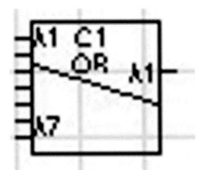

Fig. A.27. Seven-input OR gate which performs the logical OR operation

Truth Table A.27

| A | B | C | D | E | F | G | Output |
|---|---|---|---|---|---|---|---|
| $\lambda_1[0]$ | $\lambda_2[0]$ | $\lambda_3[0]$ | $\lambda_4[0]$ | $\lambda_5[0]$ | $\lambda_6[0]$ | $\lambda_7[0]$ | $\lambda_1[0]$ |
| $\lambda_1[1]$ | $\lambda_2[0]$ | $\lambda_3[0]$ | $\lambda_4[0]$ | $\lambda_5[0]$ | $\lambda_6[0]$ | $\lambda_7[0]$ | $\lambda_1[1]$ |
| .. | .. | .. | .. | .. | .. | .. | $\lambda_1[1]$ |
| .. | .. | .. | .. | .. | .. | .. | $\lambda_1[1]$ |
| $\lambda_1[1]$ | $\lambda_2[1]$ | $\lambda_3[1]$ | $\lambda_4[1]$ | $\lambda_5[1]$ | $\lambda_6[1]$ | $\lambda_7[1]$ | $\lambda_1[1]$ |

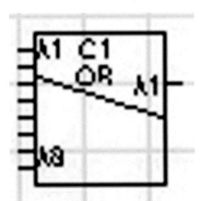

Fig. A.28. Eight-input OR gate which performs the logical OR operation

Truth Table A.28

| A | B | C | D | E | F | G | H | Output |
|---|---|---|---|---|---|---|---|---|
| $\lambda_1[0]$ | $\lambda_2[0]$ | $\lambda_3[0]$ | $\lambda_4[0]$ | $\lambda_5[0]$ | $\lambda_6[0]$ | $\lambda_7[0]$ | $\lambda_8[0]$ | $\lambda_1[0]$ |
| $\lambda_1[1]$ | $\lambda_2[0]$ | $\lambda_3[0]$ | $\lambda_4[0]$ | $\lambda_5[0]$ | $\lambda_6[0]$ | $\lambda_7[0]$ | $\lambda_8[0]$ | $\lambda_1[1]$ |
| .. | .. | .. | .. | .. | .. | .. | .. | $\lambda_1[1]$ |
| .. | .. | .. | .. | .. | .. | .. | .. | $\lambda_1[1]$ |
| $\lambda_1[1]$ | $\lambda_2[1]$ | $\lambda_3[1]$ | $\lambda_4[1]$ | $\lambda_5[1]$ | $\lambda_6[1]$ | $\lambda_7[1]$ | $\lambda_8[1]$ | $\lambda_1[1]$ |

Fig. A.29. Two-input EXOR gate which performs the logical EXOR operation

Truth Table A.29

| A | B | Output |
|---|---|---|
| $\lambda_1[0]$ | $\lambda_2[0]$ | $\lambda_1[0]$ |
| $\lambda_1[1]$ | $\lambda_2[0]$ | $\lambda_1[1]$ |
| $\lambda_1[0]$ | $\lambda_2[1]$ | $\lambda_1[1]$ |
| $\lambda_1[1]$ | $\lambda_2[1]$ | $\lambda_1[0]$ |

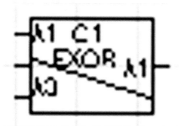

Fig. A.30. Three-input EXOR gate which performs the logical EXOR operation

Truth Table A.30

| A | B | C | Output |
|---|---|---|---|
| $\lambda_1[0]$ | $\lambda_2[0]$ | $\lambda_3[0]$ | $\lambda_1[0]$ |
| $\lambda_1[1]$ | $\lambda_2[0]$ | $\lambda_3[0]$ | $\lambda_1[1]$ |
| $\lambda_1[0]$ | $\lambda_2[1]$ | $\lambda_3[0]$ | $\lambda_1[1]$ |
| $\lambda_1[1]$ | $\lambda_2[1]$ | $\lambda_3[0]$ | $\lambda_1[0]$ |
| $\lambda_1[0]$ | $\lambda_2[0]$ | $\lambda_3[1]$ | $\lambda_1[1]$ |
| $\lambda_1[1]$ | $\lambda_2[0]$ | $\lambda_3[1]$ | $\lambda_1[0]$ |
| $\lambda_1[0]$ | $\lambda_2[1]$ | $\lambda_3[1]$ | $\lambda_1[0]$ |
| $\lambda_1[1]$ | $\lambda_2[1]$ | $\lambda_3[1]$ | $\lambda_1[0]$ |

Fig. A.31. Four-input EXOR gate which performs the logical EXOR operation

Truth Table A.31

| A | B | C | D | Output |
|---|---|---|---|---|
| $\lambda_1[0]$ | $\lambda_2[0]$ | $\lambda_3[0]$ | $\lambda_4[0]$ | $\lambda_1[0]$ |
| $\lambda_1[1]$ | $\lambda_2[0]$ | $\lambda_3[0]$ | $\lambda_4[0]$ | $\lambda_1[1]$ |
| .. | .. | .. | .. | .. |
| .. | .. | .. | .. | .. |
| $\lambda_1[1]$ | $\lambda_2[1]$ | $\lambda_3[1]$ | $\lambda_4[1]$ | $\lambda_1[0]$ |

Fig. A.32. Five-input EXOR gate which performs the logical EXOR operation

Truth Table A.32

| A | B | C | D | E | Output |
|---|---|---|---|---|---|
| $\lambda_1[0]$ | $\lambda_2[0]$ | $\lambda_3[0]$ | $\lambda_4[0]$ | $\lambda_5[0]$ | $\lambda_1[0]$ |
| $\lambda_1[1]$ | $\lambda_2[0]$ | $\lambda_3[0]$ | $\lambda_4[0]$ | $\lambda_5[0]$ | $\lambda_1[1]$ |
| .. | .. | .. | .. | .. | .. |
| .. | .. | .. | .. | .. | .. |
| $\lambda_1[1]$ | $\lambda_2[1]$ | $\lambda_3[1]$ | $\lambda_4[1]$ | $\lambda_5[1]$ | $\lambda_1[0]$ |

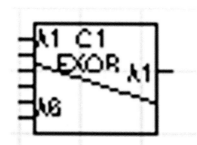

Fig. A.33. Six-input EXOR gate which performs the logical EXOR operation

Truth Table A.33

| A | B | C | D | E | F | Output |
|---|---|---|---|---|---|---|
| $\lambda_1[0]$ | $\lambda_2[0]$ | $\lambda_3[0]$ | $\lambda_4[0]$ | $\lambda_5[0]$ | $\lambda_6[0]$ | $\lambda_1[0]$ |
| $\lambda_1[1]$ | $\lambda_2[0]$ | $\lambda_3[0]$ | $\lambda_4[0]$ | $\lambda_5[0]$ | $\lambda_6[0]$ | $\lambda_1[1]$ |
| .. | .. | .. | .. | .. | .. | .. |
| .. | .. | .. | .. | .. | .. | .. |
| $\lambda_1[1]$ | $\lambda_2[1]$ | $\lambda_3[1]$ | $\lambda_4[1]$ | $\lambda_5[1]$ | $\lambda_6[1]$ | $\lambda_1[0]$ |

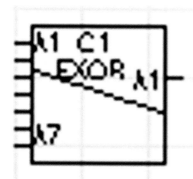

Fig. A.34. Seven-input EXOR gate which performs the logical EXOR operation

Truth Table A.34

| A | B | C | D | E | F | G | Output |
|---|---|---|---|---|---|---|---|
| $\lambda_1[0]$ | $\lambda_2[0]$ | $\lambda_3[0]$ | $\lambda_4[0]$ | $\lambda_5[0]$ | $\lambda_6[0]$ | $\lambda_7[0]$ | $\lambda_1[0]$ |
| $\lambda_1[1]$ | $\lambda_2[0]$ | $\lambda_3[0]$ | $\lambda_4[0]$ | $\lambda_5[0]$ | $\lambda_6[0]$ | $\lambda_7[0]$ | $\lambda_1[1]$ |
| .. | .. | .. | .. | .. | .. | .. | .. |
| .. | .. | .. | .. | .. | .. | .. | .. |
| $\lambda_1[1]$ | $\lambda_2[1]$ | $\lambda_3[1]$ | $\lambda_4[1]$ | $\lambda_5[1]$ | $\lambda_6[1]$ | $\lambda_7[1]$ | $\lambda_1[0]$ |

Fig. A.35. Eight-input EXOR gate which performs the logical EXOR operation

Truth Table A.35

| A | B | C | D | E | F | G | H | Output |
|---|---|---|---|---|---|---|---|--------|
| $\lambda_1[0]$ | $\lambda_2[0]$ | $\lambda_3[0]$ | $\lambda_4[0]$ | $\lambda_5[0]$ | $\lambda_6[0]$ | $\lambda_7[0]$ | $\lambda_8[0]$ | $\lambda_1[0]$ |
| $\lambda_1[1]$ | $\lambda_2[0]$ | $\lambda_3[0]$ | $\lambda_4[0]$ | $\lambda_5[0]$ | $\lambda_6[0]$ | $\lambda_7[0]$ | $\lambda_8[0]$ | $\lambda_1[1]$ |
| .. | .. | .. | .. | .. | .. | .. | .. | .. |
| .. | .. | .. | .. | .. | .. | .. | .. | .. |
| $\lambda_1[1]$ | $\lambda_2[1]$ | $\lambda_3[1]$ | $\lambda_4[1]$ | $\lambda_5[1]$ | $\lambda_6[1]$ | $\lambda_7[1]$ | $\lambda_8[1]$ | $\lambda_1[0]$ |

Fig. A.36. NOT gate which performs the logical NOT operation

Truth Table A.36

| A | Output |
|---|--------|
| $\lambda_1[0]$ | $\lambda_1[1]$ |
| $\lambda_1[1]$ | $\lambda_1[0]$ |

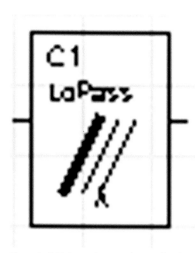

Fig. A.37. Low-Pass Filter which performs low-pass filtering

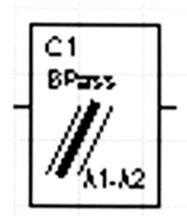

Fig. A.38. Band-Pass Filter which performs band-pass filtering

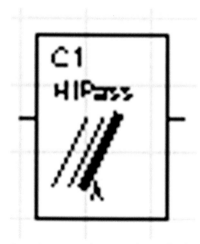

Fig. A.39. High-Pass Filter which performs high-pass filtering

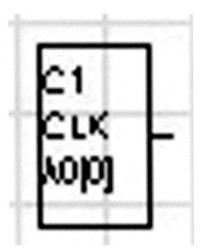

Fig. A.40. Clock used as a clock which is set to repeat low-high or high-low pulse

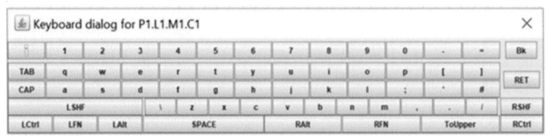

Fig. A.41. Keyboard hub which, when used, transmits an ASCII binary sequence when a key is pressed on the keyboard

| 🔥 Keyboard dialog for P1.L1.M1.C1 | | | | | | | | | | | | | | ✕ |
|---|---|---|---|---|---|---|---|---|---|---|---|---|---|---|
| ` | 1 | 2 | 3 | 4 | 5 | 6 | 7 | 8 | 9 | 0 | - | = | | Bk |
| TAB | q | w | e | r | t | y | u | i | o | p | [ | ] | | RET |
| CAP | a | s | d | f | g | h | j | k | l | ; | ' | # | | |
| LSHF | | | \ | z | x | c | v | b | n | m | , | . | / | RSHF |
| LCtrl | LFN | LAlt | | SPACE | | RAlt | | RFN | | ToUpper | | RCtrl |

Fig. A.42. This is a QWERTY keyboard locked to a certain keyboard hub. When a key is pressed the ASCII character is seen on the keyboard hub.

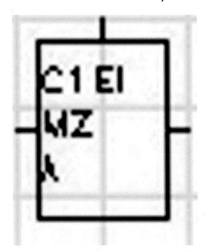

Fig. A.43. Mach-Zehnder interferometer, which is used to perform intensity modulation

170

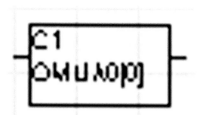

Fig. A.44. Matching unit to perform optical matching on the line to insure it is the correct length

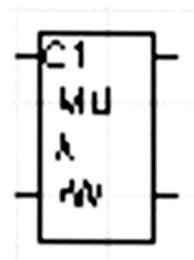

Fig. A.45. One-bit memory unit

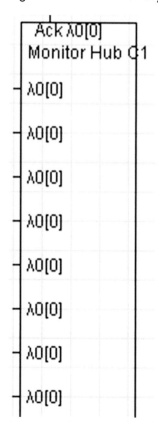

Fig. A.46. Monitor hub, which at a low level receives an ASCII character. This is displayed on the monitor connected to this hub.

171

Fig. A.47. Perform the operation of a text-based monitor updated through a back buffer at about 20Hz.

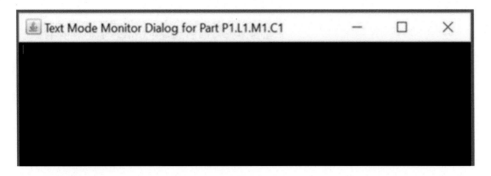

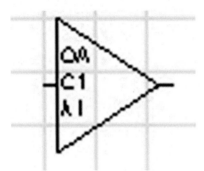

Fig. A.48. Optical amplifier, which performs optical amplification

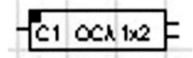

Fig. A.49. Optical coupler 1 × 2. This is a 1 to 2 optical coupler.

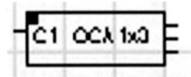

Fig. A.50. Optical coupler 1 × 3. This is a 1 to 3 optical coupler.

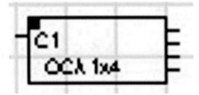

Fig. A.51. Optical coupler 1 × 4. This is a 1 to 4 optical coupler.

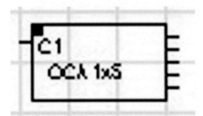

Fig. A.52. Optical coupler 1 × 5. This is a 1 to 5 optical coupler.

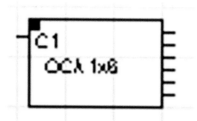

Fig. A.53. Optical coupler 1 × 6. This is a 1 to 6 optical coupler.

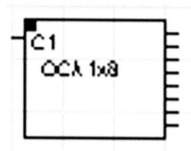

Fig. A.54. Optical coupler 1 × 8. This is a 1 to 8 optical coupler.

Fig. A.55. Optical coupler 1 × 9. This is a 1 to 9 optical coupler.

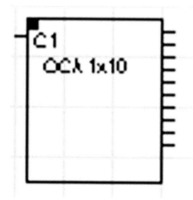

Fig. A.56. Optical coupler 1 × 10. This is a 1 to 10 optical coupler.

Fig. A.57. Optical coupler 1 × 16. This is a 1 to 16 optical coupler.

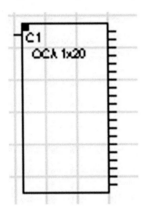

Fig. A.58. Optical coupler 1 × 20. This is a 1 to 20 optical coupler.

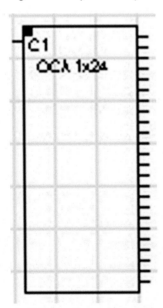

Fig. A.59. Optical coupler 1 × 24. This is a 1 to 24 optical coupler.

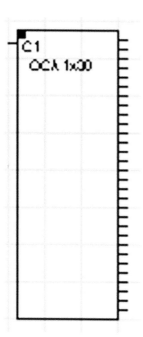

Fig. A.60. Optical coupler 1 × 30. This is a 1 to 30 optical coupler.

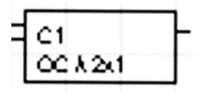

Fig. A.61. Optical coupler 2 × 1. This is a 2 to 1 optical coupler.

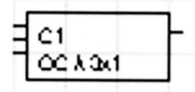

Fig. A.62. Optical coupler 3 × 1. This is a 3 to 1 optical coupler.

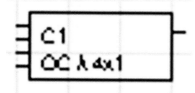

Fig. A.63. Optical coupler 4 × 1. This is a 4 to 1 optical coupler.

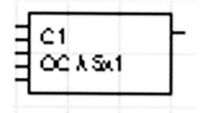

Fig. A.64. Optical coupler 5 × 1. This is a 5 to 1 optical coupler.

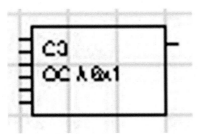

Fig. A.65. Optical coupler 6 × 1. This is a 6 to 1 optical coupler.

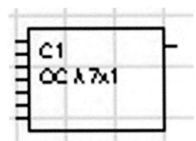

Fig. A.66. Optical coupler 7 × 1. This is a 7 to 1 optical coupler.

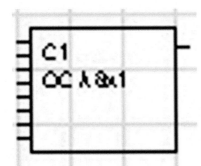

Fig. A.67. Optical coupler 8 × 1. This is a 8 to 1 optical coupler.

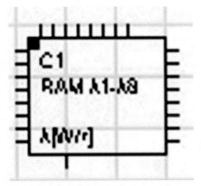

Fig. A.68. RAM8. This is a chip with 256 8-bit memory addresses.

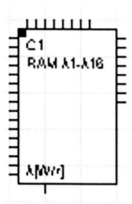

Fig. A.69. RAM16. This is a chip with 65,536 8-bit memory addresses.

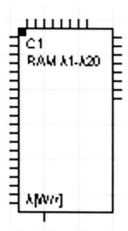

Fig. A.70. RAM20. This is a chip with 1 MB of 8-bit memory addresses.

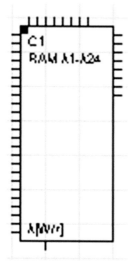

Fig. A.71. RAM24. This is a chip with 16 MB of 8-bit memory addresses.

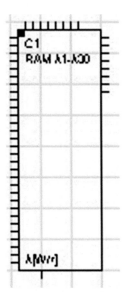

Fig. A.72. RAM30. This is a chip with 1 GB of 8-bit memory addresses.

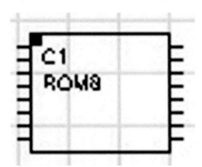

Fig. A.73. ROM8. This is a chip with 256 8-bit memory addresses.

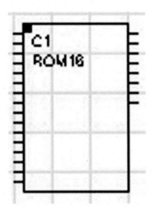

Fig. A.74. ROM16. This is a chip with 65,636 8-bit memory addresses.

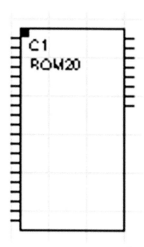

Fig. A.75. ROM20. This is a chip with 1 MB of 8-bit memory addresses.

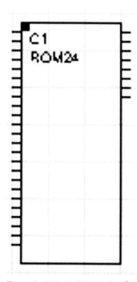

Fig. A.76. ROM24. This is a chip with 16 MB of 8-bit memory addresses.

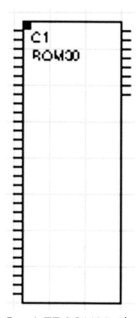

Fig. A.77. ROM30. This is a chip with 1 GB of 8-bit memory addresses.

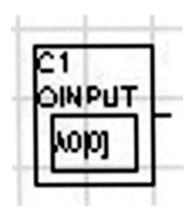

Fig. A.78. Optical input port. A laser modulated with a Mach-Zehnder interferometer to give an intensity modulated source.

Fig. A.79. Optical output port

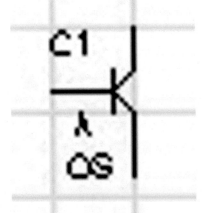

Fig. A.80. Optical switch

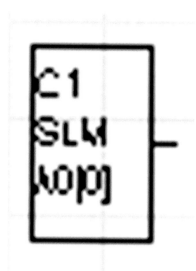

Fig. A.81. Spatial light modulator

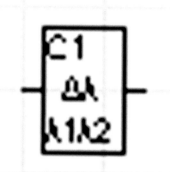

Fig. A.82. Wavelength converter

Fig. A.83. Start point same-layer intermodule link

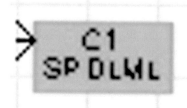

Fig. A.84. Start point different-layer intermodule link

Fig. A.85. End point same-layer intermodule link

Fig. A.86. End point different-layer intermodule link

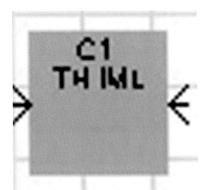

Fig. A.87. Through-hole intermodule link

Fig. A.88. SR latch

Table A.37

| Pin1 | Pin2 | Pin3 | Q |
|---|---|---|---|
| 0 | 1 | 0 | Q |
| 0 | 1 | 1 | 0 |
| 1 | 1 | 0 | 1 |

Fig. A.89. JK latch

Table A.38

| Pin1 | Pin2 | Pin3 | Q | Qt+1 |
|------|------|------|---|------|
| 0 | 1 | 0 | Q | |
| 0 | 1 | 1 | 0 | |
| 1 | 1 | 0 | 1 | |
| 1 | 1 | 1 | 1 | 0 |
| 1 | 1 | 1 | 0 | 1 |

Fig. A.90. D latch

Table A.39

| Pin1 | Pin2 | Q |
|------|------|---|
| 0 | 1 | 0 |
| 1 | 1 | 1 |

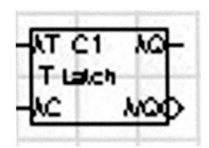

Fig. A.91. T latch

Table A.40

| Pin1 | Pin2 | Q | Qt+1 |
|------|------|---|------|
| 0 | 1 | Q | |
| 1 | 1 | 1 | 0 |
| 1 | 1 | 0 | 1 |

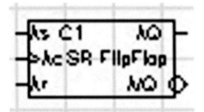

Fig. A.92. SR flip flop

Table A.41

| Pin1 | Pin2 | Pin3 | Q | Qt+1 |
|------|------|------|---|------|
| 0 | 1 | 0 | Q | |
| 0 | 1 | 1 | 1 | 0 |
| 0 | 1 | 1 | 0 | 1 |
| 1 | 1 | 0 | 1 | |

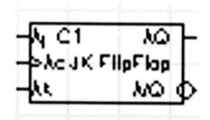

Fig. A.93. JK flip flop

Table A.42

| Pin1 | Pin2 | Pin3 | Q | Qt+1 |
|------|------|------|---|------|
| 0 | 1 | 0 | Q | |
| 0 | 0 | 0 | Q | |
| 0 | 1 | 1 | 0 | |
| 1 | 1 | 1 | 1 | 0 |
| 1 | 1 | 1 | 0 | 1 |

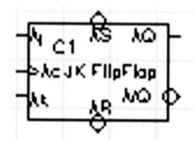

Fig. A.94. 5 Input JK flip flop

Table A.43

| Pin1 | Pin2 | Pin3 | Pin4 | Pin5 | Q | Qt+1 |
|------|------|------|------|------|---|------|
| 0 | | | | 1 | 1 | |
| 1 | | | | 0 | 0 | |
| | 0 | 1 | 1 | | 0 | |
| | 1 | 1 | 0 | | 1 | |
| | 1 | 1 | 1 | | 1 | 0 |
| | 1 | 1 | 1 | | 0 | 1 |

Fig. A.95. D flip flop

Table A.44

| Pin1 | Pin2 | Q |
|------|------|---|
| 0 | 1 | 0 |
| 1 | 1 | 1 |
| 1 | 0 | Q |
| 0 | 0 | Q |

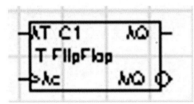

Fig. A.96. T flip flop

Table A.45

| Pin1 | Pin2 | Q | Qt+1 |
|------|------|---|------|
| 0 | 1 | Q | |
| 1 | 1 | 1 | 0 |
| 1 | 1 | 0 | 1 |

# Appendix B

This Appendix is where the code to simulate hardware algorithms is. These programs when run generate information that is specific to each hardware algorithm. The algorithms are

- Booth's algorithm for signed multiplication
- Signed division
- Floating point add/subtract
- Floating point multiply
- Floating point divide
- Tabular method and magnitude comparator

For these programs, I developed them in the NetBeans IDE which is available from netbeans.apache.org.

A generic install would be as follows:

1. Go to netbeans.apache.org.
2. Click on the downloads menu link.
3. Then click on the versions download link Apache NetBeans 11.1 and choose the installer for your operating system.
4. Download it from a nearest mirror by choosing a mirror.
5. When the downloading is finished, click on the executable file, and let it install NetBeans on your system.

# Booth's Algorithm Code Fragment

For Booth's algorithm, which is signed multiply, I set up a project as shown in Fig.B.1.

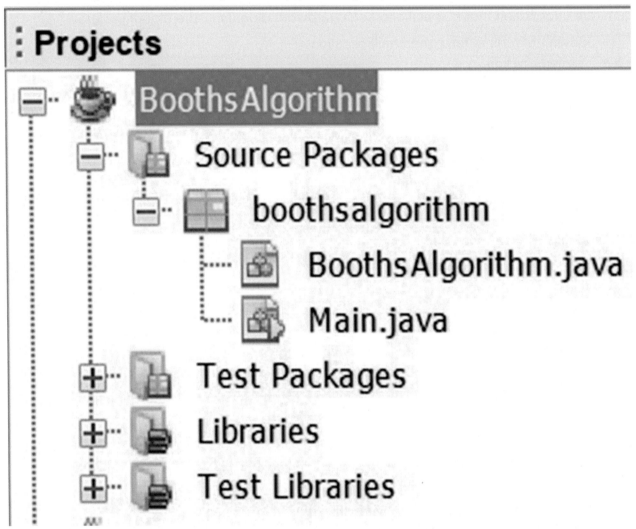

## Projects

- BoothsAlgorithm
  - Source Packages
    - boothsalgorithm
      - BoothsAlgorithm.java
      - Main.java
  - Test Packages
  - Libraries
  - Test Libraries

Fig.B.1

I created the Java project with name BoothsAlgorithm, then in this project, I created the classes BoothsAlgorithm.java and Main.java. Typed in the code and highlighted the BoothsAlgorithm project. I then chose on the menu Run and Run Project from the drop-down menu list, and it ran the code.

Fig.B.1

```
/*
 * Author: Michael Cloran 3/7/19
 * Email: michaelcloran2010@gmail.com
 *
 * Note to user. This code has absolutely no warranty. It works for me on my
 * computer and is here as is and is not deemed as fit for any particular
```

```java
 * purpose.The author cannot be held liable for damage or loss of data due
 * to its use.The author is not liable for any thing that might go wrong with
 * the codes use.
*Main.java
 */

package boothsalgorithm;

/**
 * @author mc201
 * Author: Michael Cloran 3/7/19
 * Email: michaelcloran2010@gmail.com
 * Version 1.0
 */
public class Main {
    /**
     * @param args the command line arguments
     */
    public static void main(String[] args) {
        new BoothsAlgorithm();
    }
}
/*
 * Author: Michael Cloran 3/7/19
 * Email: michaelcloran2010@gmail.com
 *
 * Note to user. This code has absolutely no warranty. It works for me on my
 * computer and is here as is and is not deemed as fit for any particular
 * purpose.The author cannot be held liable for damage or loss of data due
 * to its use.The author is not liable for any thing that might go wrong with
 * the codes use.
 */
```

```java
package boothsalgorithm;

/**
 *
 * @author mc201
 * Author: Michael Cloran 3/7/19
 * Email: michaelcloran2010@gmail.com
 * Version 1.0
 * BoothsAlgorithm.java
 */
public class BoothsAlgorithm {

    private  int[] A = new int[8];
    private  int[] Q = new int[8];
    private  int Q1;
    private  int[] M = new int[8];
    private  int RegisterWidthInBits;
    private  int count;
    private  int signBit2;


    public BoothsAlgorithm(){
        this.RegisterWidthInBits = 8;
        count = RegisterWidthInBits;
        signBit2 = 0;
                //LSB        MSB
        A = new int[] {0,0,0,0, 0,0,0,0};
        Q = new int[] {1,1,0,0, 0,0,0,0};
        Q1 = 0;
        M = new int[] {0,1,0,0, 0,0,0,0};
```

```
signBit2 = Q[RegisterWidthInBits-1] ^ M[RegisterWidthInBits-1];


printResults();//printing out the initial state of registers
System.out.println("signBit:"+signBit2);
while(count>0){
    //Going through the different Qo Q1 states
    if(Q[0] == 1 && Q1 == 0){
        System.out.println("==== Subtracting ====");
        System.out.println("Count:"+count);
        A = subtractAmM();
        arithShiftrightA_Q_Q1();
        count = count -1;
        if(count == 0) break;
    }else
    if(Q[0] == 0 && Q1 == 1){
        System.out.println("==== Adding ====");
        System.out.println("Count:"+count);
        A = addAmM();
        arithShiftrightA_Q_Q1();
        count = count -1;
        if(count ==0 )break;
    }else
    if((Q[0] == 1 && Q1 == 1) || (Q[0] == 0 && Q1 == 0)){
        System.out.println("==== Shifting ====");
        System.out.println("Count:"+count);
        arithShiftrightA_Q_Q1();
        count = count -1;
        if(count ==0 )break;
    }
}
```

```java
        if(Q[RegisterWidthInBits-1] != signBit2){
            System.out.println("signBit correction");
            setSignBit();
            printResults();
        }
    }


    public int[] subtractAmM(){
        int num1 = binaryToInteger(A);
        int num2 = binaryToInteger(M);
        byte sum = 0;


        sum = (byte) (num1 - num2);


        System.out.println("sum:"+sum+" = num1:"+num1+" - num2:"+num2);
        String binaryStr = String.format("%8s", Integer.toBinaryString(sum & 0xFF)).replace(' ',
'0');
        System.out.println("A:"+binaryStr);


        for(int i=0; i<RegisterWidthInBits; i++){
            A[(RegisterWidthInBits-1)-i] = Character.getNumericValue(binaryStr.charAt(i));
        }


        return A;
    }


    public int[] addAmM(){
        int num1 = binaryToInteger(A);
        int num2 = binaryToInteger(M);
        byte sum = (byte)(num1 + num2);
        System.out.println("sum:"+sum+" = num1:"+num1+" + num2:"+num2);
```

```java
        String binaryStr = String.format("%8s", Integer.toBinaryString(sum & 0xFF)).replace(' ',
'0');

        System.out.println("A:"+binaryStr);


        for(int i=0; i<RegisterWidthInBits ; i++){
            A[(RegisterWidthInBits-1)-i] = Character.getNumericValue(binaryStr.charAt(i));
        }


        return A;
    }


    public  void arithShiftrightA_Q_Q1(){
        System.out.println("Shifting A Q Q1 right");


        // this is the A registers width (8)+ Q registers width(8) + Q1 = 17 bits
        int[] AQQ1 = new int[17];//This is a temp register for easy shifting
        System.out.println("length of A:"+A.length);
        int signBit = A[(RegisterWidthInBits-1)];


        //copying the values into the temp register
        for(int i=16; i>=0; --i){
            if(i>RegisterWidthInBits && i<=16 ){
                AQQ1[i]=A[i-9];
            }else
            if(i<=RegisterWidthInBits && i>=1){
                AQQ1[i] = Q[i-1];
            }else
            if(i == 0){
                AQQ1[i] = Q1;
            }
        }
```

```
//doing the shift A Q Q1 right
for(int i =0; i<17; i++){
    if(i== 16){
        AQQ1[i] = signBit;


    }else{
        AQQ1[i] = AQQ1[i+1];
    }
}
//putting the values back into the registers
for(int i = 0; i<16; i++){
    if(i>RegisterWidthInBits && i<=16){
        A[i-9] = AQQ1[i];
    }else
    if(i<=RegisterWidthInBits && i>=1 ){
        Q[i-1] = AQQ1[i];
    }else
    if(i == 0){
        Q1 = AQQ1[0];
    }
}
printResults();
}


public  int binaryToInteger(int[] binaryArray){
    String str = "";
    for(int i=(RegisterWidthInBits-1) ; i>= 0; i--){
        str = str+binaryArray[i];
    }
    int decimal = Integer.parseInt(str,2);
    return decimal;
```

```java
    }

    public void setSignBit(){
        Q[(RegisterWidthInBits-1)] = signBit2;
    }


    public void printResults(){
        char uniwavelength = new Character('\u03bb');

        System.out.println("\t\tA \t\t\t\t\t\tQ \t\t\t Q1    \t\t\tM");

        for(int i=(RegisterWidthInBits-1); i>=0; i--){
            System.out.print(uniwavelength+""+(i+1)+"["+A[i]+"]");
        }
        System.out.print("\t");

        for(int i=(RegisterWidthInBits-1); i>=0; i--){
            System.out.print(uniwavelength+""+(i+1)+"["+Q[i]+"]");
        }

        System.out.print("\t"+uniwavelength+""+(1)+"["+Q1+"]\t");

        for(int i=(RegisterWidthInBits-1); i>=0; i--){
            System.out.print(uniwavelength+""+(i+1)+"["+M[i]+"]");
        }
        System.out.print("\n");
    }
}
```

# Signed Divide

For signed divide, I set up a project as shown in Fig.B.2

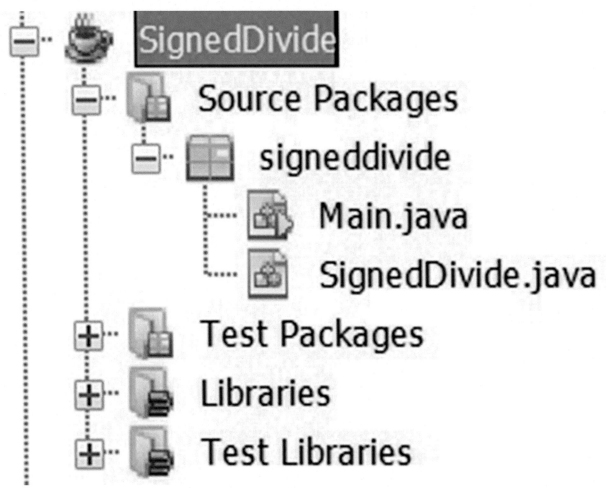

Fig.B.2

I created the Java project with name SignedDivide and then, in this project, created the classes SignedDivide.java and Main.java. I typed in the code and highlighted the SignedDivide project. Then I chose on the menu Run and Run Project from the drop-down menu list, and it ran the code.

```
/*
 * Author: Michael Cloran 6/7/19
 * Email: michaelcloran2010@gmail.com
 *
 * Note to user. This code has absolutely no warranty. It works for me on my
 * computer and is here as is and is not deemed as fit for any particular
 * purpose.The author cannot be held liable for damage or loss of data due
 * to its use.The author is not liable for any thing that might go wrong with
 * the codes use.
```

```
 */
package signeddivide;


/**

 *

 * @author mc201
 * Author: Michael Cloran 19/7/19
 * Email: michaelcloran2010@gmail.com
 * Version 1.0
*Main.java
 */
public class Main {
    public static void main(String[] args) {
        new SignedDivide();
    }
}


/*
 * Author: Michael Cloran 6/7/19
 * Email: michaelcloran2010@gmail.com
 *
 * Note to user. This code has absolutely no warranty. It works for me on my
 * computer and is here as is and is not deemed as fit for any particular
 * purpose.The author cannot be held liable for damage or loss of data due
 * to its use.The author is not liable for any thing that might go wrong with
 * the codes use.
 */
package signeddivide;


/**

 *
```

```java
 * @author mc201
 * Author: Michael Cloran 19/7/19
 * Email: michaelcloran2010@gmail.com
 * Version 1.0
 * SignedDivide.java
 */
public class SignedDivide {
                            //LSB          MSB
    private static int[] A;
    private static int[] Q;


    private static int[] M;
    private static int RegisterWidthInBits;
    private static int count;
    private static int signBit;
    /**
     * @param args the command line arguments
     */
    public SignedDivide() {
        this.A = new int[] {0,0,0,0, 0,0,0,0};
        this.Q = new int[] {1,1,1,0, 0,0,0,0};
        this.M = new int[] {0,1,0,0, 0,0,0,0};

        this.RegisterWidthInBits = 8;
        this.count = this.RegisterWidthInBits;
        this.signBit = Q[this.RegisterWidthInBits-1] ^ M[this.RegisterWidthInBits-1];

        //Adjusting to do unsigned division where the sign will
        //be added back at the end
        if(this.Q[this.RegisterWidthInBits-1]==1){
            this.Q[this.RegisterWidthInBits-1] = 0;
```

```
    }
    if(this.M[this.RegisterWidthInBits-1]==1){
        this.M[this.RegisterWidthInBits-1] = 0;
    }


    printResults();//printing out the initial state of registers

    while(this.count>0){
        System.out.println("Count:"+this.count);
        System.out.println("==== shifting AQ left ====");


        arithShiftLefttA_Q();


        System.out.println("==== subtracting ====");
        this.A = subtractAmM();
        //Testing for sign bit negative and thus below 0
        if(this.A[this.RegisterWidthInBits-1]==1){
            this.Q[0] = 0;
            System.out.println("==== adding ====");
            this.A = addAmM();
        }else{
            this.Q[0] = 1;
        }


        printResults();
        this.count = this.count - 1;
    }
    if(this.Q[this.RegisterWidthInBits-1] != this.signBit){
        this.Q[this.RegisterWidthInBits-1] = this.signBit;
        this.A[this.RegisterWidthInBits-1] = this.signBit;
        System.out.println("==== Adjusting value to show sign in AQ ====");
```

```java
        printResults();

    }

  }


  public int[] subtractAmM(){
      int num1 = binaryToInteger(this.A);
      int num2 = binaryToInteger(this.M);
      int sum = 0;


      sum = (num1 - num2);
      System.out.println("sum:"+sum+" = num1:"+num1+" - num2:"+num2);
      String binaryStr = String.format("%8s", Integer.toBinaryString(sum & 0xFF)).replace(' ',
'0');
      System.out.println("A:"+binaryStr);


      for(int i=0; i<this.RegisterWidthInBits; i++){
          this.A[(this.RegisterWidthInBits-1)-i] = Character.getNumericValue(binaryStr.
charAt(i));
      }


      return this.A;
  }


  public int[] addAmM(){
      int num1 = binaryToInteger(this.A);
      int num2 = binaryToInteger(this.M);
      int sum = num1 + num2;
      System.out.println("sum:"+sum+" = num1:"+num1+" + num2:"+num2);
      String binaryStr = String.format("%8s", Integer.toBinaryString(sum & 0xFF)).replace(' ',
'0');
      System.out.println("A:"+binaryStr+"binaryStringLength:"+binaryStr.length());
```

```java
    for(int i=0; i<this.RegisterWidthInBits ; i++){
        this.A[(this.RegisterWidthInBits-1)-i] = Character.getNumericValue(binaryStr.
charAt(i));
    }


    return this.A;
}


public void arithShiftLefttA_Q(){
    System.out.println("Shifting A Q left");

    // this is the A registers width (8)+ Q registers width(8) = 16 bits
    int[] AQ = new int[16];//This is a temp register for easy shifting
    System.out.println("length of A:"+this.A.length);
    int ctr=1;

    //copying the values into the temp register
    for(int i=15; i>=0; --i){
        if(i>=this.RegisterWidthInBits && i<=15 ){
            AQ[i]=this.A[i-8];
        }else
        if(i<this.RegisterWidthInBits && i>=0){
            AQ[i] = this.Q[i];
        }
    }
    //doing the shift A Q left
    for(int i =15; i>=0; i--){
        if(i== 0){
            AQ[i] = 0;

        }else{
```

```java
            AQ[i] = AQ[i-1];
        }
    }
    //putting the values back into the registers
    for(int i = 0; i<16; i++){
        if(i>=this.RegisterWidthInBits && i<16){
            this.A[i-8] = AQ[i];
        }else
        if(i<this.RegisterWidthInBits && i>=0 ){
            this.Q[i] = AQ[i];
        }
    }
}


public int binaryToInteger(int[] binaryArray){
    String str = "";
    for(int i=(this.RegisterWidthInBits-1) ; i>= 0; i-){
        str = str+binaryArray[i];
    }

    int decimal = 0;
    for(int i = (this.RegisterWidthInBits-1); i>=0; i-){
        decimal = decimal + (int)binaryArray[i]*(int)Math.pow(2, i);
    }
    return decimal;
}


public void printResults(){
    char uniwavelength = new Character('\u03bb');
    System.out.println("\t\tA \t\t\t\t\t\tQ   \t\t\t\t\tM");
```

```java
        for(int i=(this.RegisterWidthInBits-1); i>=0; i--){
            System.out.print(uniwavelength+""+(i+1)+"["+this.A[i]+"]");
        }
        System.out.print("\t");
        for(int i=(this.RegisterWidthInBits-1); i>=0; i--){
            System.out.print(uniwavelength+""+(i+1)+"["+this.Q[i]+"]");
        }
        System.out.print("\t");
        for(int i=(this.RegisterWidthInBits-1); i>=0; i--){
            System.out.print(uniwavelength+""+(i+1)+"["+this.M[i]+"]");
        }
        System.out.print("\n");
    }
}
```

# Floating point Add/Subtract

For floating point add or subtract, I set up a project as shown in Fig.B.3

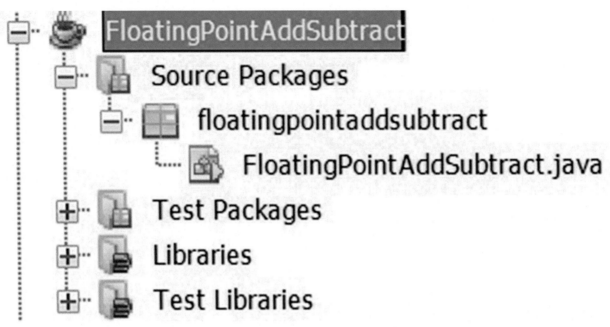

Fig.B.3

I created the Java project with name FloatingPointAddSubtract, and then in this project, I created the classes FloatingPointAddSubtract.java. I typed in the code and highlighted the FloatingPointAddSubtract project. I then chose on the menu Run and Run Project from the drop-down menu list, and it ran the code.

```
/*
 * Author: Michael Cloran 3/7/19
 * Email: michaelcloran2010@gmail.com
 *
 * Note to user. This code has absolutely no warranty. It works for me on my
 * computer and is here as is and is not deemed as fit for any particular
 * purpose.The author cannot be held liable for damage or loss of data due
 * to its use.The author is not liable for any thing that might go wrong with
 * the codes use.
 */
package floatingpointaddsubtract;
```

```java
/**
 *
 * @author mc201
 * Author: Michael Cloran 3/7/19
 * Email: michaelcloran2010@gmail.com
 * Version 1.0
 * FloatingPointAddSubtract.java
 */
public class FloatingPointAddSubtract {

    private static int RegisterWidthInBits = 32;
    private static int exponentWidthInBits = 8;
    private static int significandWidthInBits = 23;

    private static int signbitX = 0;
    private static int[] biasedExponentX = new int[] {1,1,0,0, 0,0,0,0};
    private static int[] significandX = new int[]{0,1,0,0 ,0,0,0,0, 0,0,0,0, 0,0,0,0, 0,0,0,0, 0,0,0};

    private static int signbitY = 0;
    private static int[] biasedExponentY = new int[] {0,1,0,0, 0,0,0,0};
    private static int[] significandY = new int[]{1,1,1,0 ,0,0,0,0, 0,0,0,0, 0,0,0,0, 0,0,0,0, 0,0,0};

    private static int signbitZ = 0;
    private static int[] biasedExponentZ = new int[] {0,0,0,0, 0,0,0,0};
    private static int[] significandZ = new int[]{0,0,0,0 ,0,0,0,0, 0,0,0,0, 0,0,0,0, 0,0,0,0, 0,0,0};
    /**
     * @param args the command line arguments
     */
    public static void main(String[] args) {

        //doing subtract uncomment these lines to do addition
        //signbitY = 1;
        //signbitZ = signbitY;
```

```
//end setup for subtract


if(significandEqualZero(biasedExponentX)==true){
    for(int i=0; i<significandWidthInBits;i++ ){
        significandZ[i] =  significandY[i];
    }
}else{
    if(significandEqualZero(biasedExponentY)==true){
        for(int i=0; i<significandWidthInBits;i++ ){
            significandZ[i] =  significandX[i];
        }
    }else{
        alignSignificands();
    }
}
}


public static boolean significandEqualZero(int[] significand){
    for(int i=0; i<significandWidthInBits; i++ ){
        if(significand[i] !=0){
            return false;
        }
    }
    return true;
}


public static boolean exponentsEqual(){
    for(int i=0; i<exponentWidthInBits; i++){
        if(biasedExponentX[i] != biasedExponentY[i]){
            return false;
        }
    }
    return true;
```

```java
    }

    public static void alignSignificands(){
        int A = 0;
        int B = 0;
        String binaryStr = "";

        A = binaryToInteger(biasedExponentX);
        B = binaryToInteger(biasedExponentY);
        if(exponentsEqual()==true){
            AddSubSignedSignificands();
        }else{
            if(A < B){

                A = A +1;
                binaryStr = String.format("%8s", Integer.toBinaryString(A & 0xFF)).replace(' ' , '0');
                for(int i=0; i<exponentWidthInBits; i++){
                    biasedExponentX[(exponentWidthInBits-1)-i] = Character.getNumericValue(binaryStr.charAt(i));
                }
                //shifting
                for(int i=significandWidthInBits-1; i>=0; i-- ){
                    if(i== 0){
                        significandX[i]=0;
                    }else{
                        significandX[i] = significandX[i-1];
                    }
                }

                if(significandEqualZero(significandX)==true){
                    signbitZ = signbitY;
                    for(int i=0; i< exponentWidthInBits; i++){
                        biasedExponentZ[i] = biasedExponentY[i] ;
```

```
        }
        for(int i=0; i< significandWidthInBits; i++){
            significandZ[i] = significandY[i];
        }
        printResults();
        return;
    }else{
        System.out.println("Aligning Significands");
        alignSignificands();
    }

}else{

    B = B +1;
    binaryStr = String.format("%8s", Integer.toBinaryString(B & 0xFF)).replace(' ' , '0');
    for(int i=0; i<exponentWidthInBits; i++){
        biasedExponentY[(exponentWidthInBits-1)-i] = Character.getNumericValue(binaryStr.charAt(i));
    }
    //shifting
    for(int i=significandWidthInBits-1; i>=0; i-- ){
        if(i== 0){
            significandY[i]=0;
        }else{
            significandY[i] = significandY[i-1];
        }
    }

    if(significandEqualZero(significandY)==true){
        signbitZ = signbitX;
        for(int i=0; i< exponentWidthInBits; i++){
            biasedExponentZ[i] = biasedExponentX[i] ;
        }
```

```java
                for(int i=0; i< significandWidthInBits; i++){
                    significandZ[i] = significandX[i];
                }
                printResults();
                return;
            }else{
                System.out.println("Aligning Significands");
                alignSignificands();
            }
        }
    }
}


public static void AddSubSignedSignificands(){
    int A = binaryToInteger(significandX);
    int B = binaryToInteger(significandY);
    int Z = 0;
    if(signbitX ==1 && signbitY ==0 ){
        Z = -A+B;
    }else
    if(signbitX ==0 && signbitY ==1 ){
        Z = A-B;
    }else
    if(signbitX ==1 && signbitY ==1 ){
        Z = -A -B;
    }else{
        Z = A+B;
    }

    if(Z == 0){
        for(int i=0; i<exponentWidthInBits;i++ ){
            biasedExponentZ[i] = 0;
        }
```

```java
        for(int i=0; i<significandWidthInBits;i++ ){
            significandZ[i] = 0;
        }
        printResults();
        return;
    }else{

        if(Z < 0){
            signbitZ = 1;
            Z = Z * -1;
        }
        for(int i=0; i<exponentWidthInBits;i++ ){
            biasedExponentZ[i] = biasedExponentX[i];
        }


        String binaryStr = String.format("%23s", Integer.toBinaryString(Z )).replace(' ' , '0');
        for(int i=0; i<significandWidthInBits; i++){
            significandZ[(significandWidthInBits-1)-i] = Character.getNumericValue(binaryStr.
charAt(i));
        }

        if(signbitZ == 1){
            System.out.println(" Z:"+(Z*-1)+" A:"+A+" B:"+B);
        }else{
            System.out.println(" Z:"+Z+" A:"+A+" B:"+B);
        }

        if(Z>8388607){
            //shifting
            for(int i=0; i<significandWidthInBits; i++ ){
                if(i== significandWidthInBits-1){
                    significandX[i]=0;
                }else{
```

```java
                significandX[i] = significandX[i+1];
            }
        }


        int exponent = binaryToInteger(biasedExponentX);


        exponent = exponent +1;


        if(exponent>255){
            System.out.print("Exponent overflow");
            printResults();
            return;
        }else{
            normalizeResult();
        }
    }else{
        if(Z < 0){
            signbitZ = 1;
        }
        System.out.println("normalising results");
        normalizeResult();
    }
}
}


public static void normalizeResult(){
    int tempArray[] = new int[32];
    tempArray[31] = signbitZ;
    for(int i =0; i<exponentWidthInBits;i++ ){
        tempArray[i+23] = biasedExponentZ[i];
    }
    for(int i=0; i<significandWidthInBits-1;i++ ){
        tempArray[i] = significandZ[i];
```

```java
}

signbitZ = tempArray[31];

System.out.println();
for(int i = (tempArray.length-2); i>=23;i-- ){
    biasedExponentZ[i-23] = tempArray[i];
}
for(int i=0; i<significandWidthInBits-1;i++ ){
    significandZ[i] = tempArray[i];
}

if((signbitZ) == 1){
    //Todo Round result needed
    System.out.println("Todo round needed");
    printResults();
}else{
    //shifting
    for(int i =tempArray.length-1; i>=0; i--){
        if(i== 0){
            tempArray[i] = 0;

        }else{
            tempArray[i] = tempArray[i-1];
        }
    }

    signbitZ = tempArray[31];
    for(int i =exponentWidthInBits-1; i>23;i-- ){
        biasedExponentZ[i-23] = tempArray[i];
    }
    for(int i=0; i<significandWidthInBits;i++ ){
        significandZ[i] = tempArray[i];
```

```
        }

        int tempValue = binaryToInteger(biasedExponentZ);
        tempValue = tempValue -1;


        String binaryStr = String.format("%8s", Integer.toBinaryString(tempValue & 0xFF)).
replace(' ' , '0');
        for(int i=0; i<exponentWidthInBits; i++){
            biasedExponentZ[(exponentWidthInBits-1)-i] = Character.getNumericValue(binaryS-
tr.charAt(i));
        }


        if((binaryToInteger(biasedExponentZ)<= 0 &&  signbitZ == 1) || (binaryToInteger(bia-
sedExponentZ)<= 0 &&  signbitZ == 0)){
            System.out.println("exponent underflow");
            printResults();
            return;
        }else{
            System.out.println("normalising results");
            normalizeResult();
        }
    }
}


public static int binaryToInteger(int[] binaryArray){
    String str = "";
    for(int i=(binaryArray.length-1) ; i>= 0; i--){
        str = str+binaryArray[i];
    }

    int decimal = 0;
    for(int i = (binaryArray.length-1); i>=0; i--){
        decimal = decimal + (int)binaryArray[i]*(int)Math.pow(2, i);
    }
```

```java
        return decimal;
    }


    public static void printResults(){
        char uniwavelength = new Character('\u03bb');
        System.out.println("\n\t\tSign \t\t\tEX \t\t\t\t\t\t\t\t\t S");
        System.out.print("InputX \t");
        System.out.print(uniwavelength+""+(32)+"["+signbitX+"]");
        System.out.print("\t");
        for(int i=exponentWidthInBits-1; i>=0;i-- ){
            System.out.print(uniwavelength+""+(32-exponentWidthInBits+i)+"["+biasedExponentX-
[i]+"]");
        }
        System.out.print("\t");
        for(int i=significandWidthInBits-1; i>=0;i-- ){
            System.out.print(uniwavelength+""+(32-exponentWidthInBits-significandWidthIn-
Bits+i)+"["+significandX[i]+"]");
        }


        System.out.println();
        System.out.print("InputY \t");
        System.out.print(uniwavelength+""+(32)+"["+signbitY+"]");
        System.out.print("\t");
        for(int i=exponentWidthInBits-1; i>=0;i-- ){
            System.out.print(uniwavelength+""+(32-exponentWidthInBits+i)+"["+biasedExponen-
tY[i]+"]");
        }
        System.out.print("\t");
        for(int i=significandWidthInBits-1; i>=0;i-- ){
            System.out.print(uniwavelength+""+(32-exponentWidthInBits-significandWidthIn-
Bits+i)+"["+significandY[i]+"]");
        }


        System.out.println();
```

```java
System.out.print("ResultZ\t");
System.out.print(uniwavelength+""+(32)+"["+signbitZ+"]");
System.out.print("\t");
for(int i=exponentWidthInBits-1; i>=0;i-- ){
    System.out.print(uniwavelength+""+(32-exponentWidthInBits+i)+"["+biasedExpo-
nentZ[i]+"]");
}
System.out.print("\t");
for(int i=significandWidthInBits-1; i>=0;i-- ){
    System.out.print(uniwavelength+""+(32-exponentWidthInBits-significandWidthIn-
Bits+i)+"["+significandZ[i]+"]");
}
System.out.println();
    }
}
```

# Floating Point Signed Multiply

For floating point multiply, I set up a project as shown in Fig.B.4

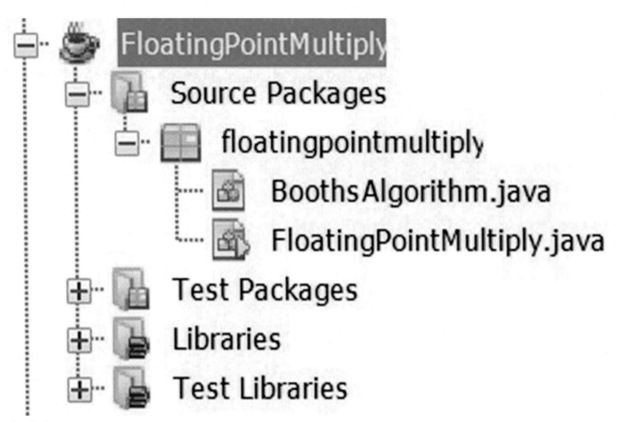

Fig.B.4

I created the Java project with name FloatingPointMultiply and then, in this project, created the classes FloatingPointMultiply.java and BoothsAlgorithm.java. I typed in the code and highlighted the FloatingPointMultiply project. Then I chose on the menu Run and Run Project from the drop-down menu list, and it ran the code.

```
/*
 * Author: Michael Cloran 3/7/19
 * Email: michaelcloran2010@gmail.com
 *
 * Note to user. This code has absolutely no warranty. It works for me on my
 * computer and is here as is and is not deemed as fit for any particular
 * purpose.The author cannot be held liable for damage or loss of data due
 * to its use.The author is not liable for any thing that might go wrong with
 * the codes use.
```

```java
    */
package floatingpointmultiply;
/**
 *
 * @author mc201
 * Author: Michael Cloran 3/7/19
 * Email: michaelcloran2010@gmail.com
 * Version 1.0
 * FloatingPointMultiply.java
 */
public class FloatingPointMultiply {

    private static int RegisterWidthInBits = 32;
    private static int exponentWidthInBits = 8;
    private static int significandWidthInBits = 23;

    private static int Q1 = 0;
    private static int[] A = new int[]{0,0,0,0 ,0,0,0,0, 0,0,0,0, 0,0,0,0, 0,0,0,0, 0,0,0};

    private static int signbitX = 1;
    private static int[] biasedExponentX = new int[] {0,0,0,0, 0,0,0,1};
    private static int[] significandX = new int[]{0,1,0,0 ,0,0,0,0, 0,0,0,0, 0,0,0,0, 0,0,0,0, 0,0,0};

    private static int signbitY = 0;
    private static int[] biasedExponentY = new int[] {1,1,0,0, 0,0,0,0};
    private static int[] significandY = new int[]{1,1,1,0 ,0,0,0,0, 0,0,0,0, 0,0,0,0, 0,0,0,0, 0,0,0};

    private static int signbitZ = 0;
    private static int[] biasedExponentZ = new int[] {0,0,0,0, 0,0,0,0};
    private static int[] significandZ = new int[]{0,0,0,0 ,0,0,0,0, 0,0,0,0, 0,0,0,0, 0,0,0,0, 0,0,0};

    private static int signBit2 = signbitX ^ signbitY;
    /**
```

```java
     * @param args the command line arguments
     */
    public static void main(String[] args) {
        if(arrayValuiesEqualsZero(biasedExponentX)==true && arrayValuiesEqualsZero(signifi-
candX)==true && signbitX==0){
            signbitZ = 0;
            for(int i=0; i<exponentWidthInBits;i++ ){
                biasedExponentZ[i] =  0;
            }
            for(int i=0; i<significandWidthInBits;i++ ){
                significandZ[i] =  0;
            }
            printFinalResults();
            return;
        }else{
            if(arrayValuiesEqualsZero(biasedExponentY)==true && arrayValuiesEqualsZero(signifi-
candY)==true && signbitY==0){
                signbitZ = 0;
                for(int i=0; i<exponentWidthInBits;i++ ){
                    biasedExponentZ[i] =  0;
                }
                for(int i=0; i<significandWidthInBits;i++ ){
                    significandZ[i] =  0;
                }
                printFinalResults();
                return;
            }else{
                byte exponentX = (byte)binaryToInteger(biasedExponentX);
                byte exponentY = (byte)binaryToInteger(biasedExponentY);
                byte exponentZ =(byte)(exponentX + exponentY);

                exponentZ = (byte)(exponentZ - 127);//biased exponent
```

```java
        String binaryStr = String.format("%8s", Integer.toBinaryString(exponentZ & 0xFF)).
replace(' ' , '0');
        for(int i=0; i<exponentWidthInBits; i++){
            biasedExponentZ[(exponentWidthInBits-1)-i] = Character.getNumericValue(bina-
ryStr.charAt(i));
        }
        if(exponentZ>255){
            System.out.println("Exponent overflow");
            return;
        }else{
            if(exponentZ<= -255){
                System.out.print("Exponent underflow");
                return;
            }else{

                signbitZ = signbitX;
                for(int i=0; i<significandWidthInBits; i++){
                    significandZ[i] = significandX[i];
                }


                //multiply Significands
                BoothsAlgorithm baObj = new BoothsAlgorithm(A,significandZ,significandY);
                int[] temp = baObj.getQ();
                for(int i=0; i<significandWidthInBits; i++){
                    significandZ[i] = temp[i];
                }


                normalizeResult();
                System.out.println("Todo round result");


                if(signbitZ != signBit2){
                    System.out.println("signBit correction");
                    setSignBit();
```

```java
                printFinalResults();
            }
            return;
        }
      }
    }
  }
}


public static void setSignBit(){
    signbitZ = signBit2;
}


public static int binaryToInteger(int[] binaryArray){
    String str = "";
    for(int i=(binaryArray.length-1) ; i>= 0; i--){
      str = str+binaryArray[i];
    }


    int decimal = 0;
    for(int i = (binaryArray.length-1); i>=0; i--){
       decimal = decimal + (int)binaryArray[i]*(int)Math.pow(2, i);
    }
    return decimal;
}


public static boolean significandEqualZero(int[] significand){
    for(int i=0; i<significandWidthInBits; i++ ){
      if(significand[i] !=0){
          return false;
      }
    }
    return true;
```

```java
}

public static boolean arrayValuiesEqualsZero(int[] arrayOfValues){
    for(int i=0; i<arrayOfValues.length; i++ ){
        if(arrayOfValues[i] !=0){
            return false;
        }
    }
    return true;
}

public static void normalizeResult(){
    int tempArray[] = new int[32];
    tempArray[31] = signbitZ;
    for(int i =0; i<exponentWidthInBits;i++ ){
        tempArray[i+23] = biasedExponentZ[i];
    }
    for(int i=0; i<significandWidthInBits-1;i++ ){
        tempArray[i] = significandZ[i];
    }

    signbitZ = tempArray[31];

    System.out.println();
    for(int i = (tempArray.length-2); i>=23;i-- ){
        biasedExponentZ[i-23] = tempArray[i];
    }
    for(int i=0; i<significandWidthInBits-1;i++ ){
        significandZ[i] = tempArray[i];
    }

    if((signbitZ) == 1){
        //Todo Round result needed
```

```java
            System.out.println("Todo round result");
            printFinalResults();
        }else{
            //shifting
            for(int i =tempArray.length-1; i>=0; i--){
                if(i== 0){
                    tempArray[i] = 0;

                }else{
                    tempArray[i] = tempArray[i-1];
                }
            }


            signbitZ = tempArray[31];
            for(int i =exponentWidthInBits-1; i>23;i-- ){
                biasedExponentZ[i-23] = tempArray[i];
            }
            for(int i=0; i<significandWidthInBits;i++ ){
                significandZ[i] = tempArray[i];
            }


            int tempValue = binaryToInteger(biasedExponentZ);
            tempValue = tempValue -1;


            String binaryStr = String.format("%8s", Integer.toBinaryString(tempValue & 0xFF)).
replace(' ' , '0');
            for(int i=0; i<exponentWidthInBits; i++){
                biasedExponentZ[(exponentWidthInBits-1)-i] = Character.getNumericValue(binaryS-
tr.charAt(i));
            }


            if((binaryToInteger(biasedExponentZ)<= 0 &&  signbitZ == 1) || (binaryToInteger(bia-
sedExponentZ)<= 0 &&  signbitZ == 0)){
                System.out.println("exponent underflow");
```

```java
        printFinalResults();
        return;
      }else{
        System.out.println("normalising results");
        normalizeResult();
      }
    }
  }


public static void printFinalResults(){
    char uniwavelength = new Character('\u03bb');
    System.out.println("\n\t\tSign \t\t\tEX \t\t\t\t\t\t\t\t S");
    System.out.print("InputX \t");
    System.out.print(uniwavelength+""+(32)+"["+signbitX+"]");
    System.out.print("\t");
    for(int i=exponentWidthInBits-1; i>=0;i-- ){
        System.out.print(uniwavelength+""+(32-exponentWidthInBits+i)+"["+biasedExponentX-
[i]+"]");
    }
    System.out.print("\t");
    for(int i=significandWidthInBits-1; i>=0;i-- ){
        System.out.print(uniwavelength+""+(32-exponentWidthInBits-significandWidthIn-
Bits+i)+"["+significandX[i]+"]");
    }
    System.out.println("biased form");

    System.out.println();
    System.out.print("InputY \t");
    System.out.print(uniwavelength+""+(32)+"["+signbitY+"]");
    System.out.print("\t");
    for(int i=exponentWidthInBits-1; i>=0;i-- ){
        System.out.print(uniwavelength+""+(32-exponentWidthInBits+i)+"["+biasedExponen-
tY[i]+"]");
    }
```

```java
        System.out.print("\t");
        for(int i=significandWidthInBits-1; i>=0;i-- ){
            System.out.print(uniwavelength+""+(32-exponentWidthInBits-significandWidth-
Bits+i)+"["+significandY[i]+"]");
        }
        System.out.println("biased form");


        System.out.println();
        System.out.print("ResultZ\t");
        System.out.print(uniwavelength+""+(32)+"["+signbitZ+"]");
        System.out.print("\t");
        for(int i=exponentWidthInBits-1; i>=0;i-- ){
            System.out.print(uniwavelength+""+(32-exponentWidthInBits+i)+"["+biasedExpo-
nentZ[i]+"]");
        }
        System.out.print("\t");
        for(int i=significandWidthInBits-1; i>=0;i-- ){
            System.out.print(uniwavelength+""+(32-exponentWidthInBits-significandWidthIn-
Bits+i)+"["+significandZ[i]+"]");
        }
        System.out.println("unbiased form");
    }
}


/*
 * Author: Michael Cloran 3/7/19
 * Email: michaelcloran2010@gmail.com
 *
 * Note to user. This code has absolutely no warranty. It works for me on my
 * computer and is here as is and is not deemed as fit for any particular
 * purpose.The author cannot be held liable for damage or loss of data due
 * to its use.The author is not liable for any thing that might go wrong with
 * the codes use.
 */
```

```java
package floatingpointmultiply;

/**
 *
 * @author mc201
 * Author: Michael Cloran 3/7/19
 * Email: michaelcloran2010@gmail.com
 * Version 1.0
 * BoothsAlgorithm.java
 */
public class BoothsAlgorithm {

    private  int[] A;
    private  int[] Q;
    private  int Q1;
    private  int[] M;
    private  int RegisterWidthInBits;
    private  int count;
    private  int signBit2;


    public BoothsAlgorithm(int[] A,int[] Q,int[] M){

        this.A = A;
        this.Q = Q;
        this.M = M;
        this.RegisterWidthInBits = A.length;
        count = RegisterWidthInBits;
        signBit2 = 0;

        printResults();//printing out the initial state of registers
        System.out.println("signBit:"+signBit2);
        while(count>0){
```

```java
        //Going through the different Qo Q1 states
        if(Q[0] == 1 && Q1 == 0){
            System.out.println("==== Subtracting ====");
            System.out.println("Count:"+count);
            A = subtractAmM();
            arithShiftrightA_Q_Q1();
            count = count -1;
            if(count == 0) break;
        }else
        if(Q[0] == 0 && Q1 == 1){
            System.out.println("==== Adding ====");
            System.out.println("Count:"+count);
            A = addAmM();
            arithShiftrightA_Q_Q1();
            count = count -1;
            if(count ==0 )break;
        }else
        if((Q[0] == 1 && Q1 == 1) || (Q[0] == 0 && Q1 == 0)){
            System.out.println("==== Shifting ====");
            System.out.println("Count:"+count);
            arithShiftrightA_Q_Q1();
            count = count -1;
            if(count ==0 )break;
        }
    }
}


public  int[] subtractAmM(){
    int num1 = binaryToInteger(A);
    int num2 = binaryToInteger(M);
    int sum = 0;

    sum = (int) (num1 - num2);
```

```java
    System.out.println("sum:"+sum+" = num1:"+num1+" - num2:"+num2);
    String binaryStr = String.format("%23s", Integer.toBinaryString(sum)).replace(' ' , '0');
    System.out.println("A:"+binaryStr);


    for(int i=0; i<RegisterWidthInBits; i++){
        A[(RegisterWidthInBits-1)-i] = Character.getNumericValue(binaryStr.charAt(i));
    }


    return A;
}


public  int[] addAmM(){
    int num1 = binaryToInteger(A);
    int num2 = binaryToInteger(M);
    int sum = (int)(num1 + num2);
    System.out.println("sum:"+sum+" = num1:"+num1+" + num2:"+num2);
    String binaryStr = String.format("%23s", Integer.toBinaryString(sum)).replace(' ', '0');
    System.out.println("A:"+binaryStr);


    for(int i=0; i<RegisterWidthInBits ; i++){
        A[(RegisterWidthInBits-1)-i] = Character.getNumericValue(binaryStr.charAt(i));
    }


    return A;
}


public  void arithShiftrightA_Q_Q1(){
    System.out.println("Shifting A Q Q1 right");


    // this is the A registers width (23)+ Q registers width(23) + Q1 = 17 bits
    int[] AQQ1 = new int[47];//This is a temp register for easy shifting
    System.out.println("length of A:"+A.length);
```

```
int signBit = A[(RegisterWidthInBits-1)];


//copying the values into the temp register
for(int i=(AQQ1.length-1); i>=0; --i){
    if(i>RegisterWidthInBits && i<=(AQQ1.length-1) ){
        AQQ1[i]=A[i-(RegisterWidthInBits+1)];
    }else
    if(i<=RegisterWidthInBits && i>=1){
        AQQ1[i] = Q[i-1];
    }else
    if(i == 0){
        AQQ1[i] = Q1;
    }
}
//doing the shift A Q Q1 right
for(int i =0; i<AQQ1.length; i++){
    if(i== (AQQ1.length-1)){
        AQQ1[i] = signBit;


    }else{
        AQQ1[i] = AQQ1[i+1];
    }
}
//putting the values back into the registers
for(int i = 0; i<(AQQ1.length-1); i++){
    if(i>RegisterWidthInBits && i<=(AQQ1.length-1)){
        A[i-(RegisterWidthInBits+1)] = AQQ1[i];
    }else
    if(i<=RegisterWidthInBits && i>=1 ){
        Q[i-1] = AQQ1[i];
    }else
    if(i == 0){
        Q1 = AQQ1[0];
```

```java
        }
    }
    printResults();
}


public  int binaryToInteger(int[] binaryArray){
    String str = "";
    for(int i=(RegisterWidthInBits-1) ; i>= 0; i--){
        str = str+binaryArray[i];
    }
    int decimal = Integer.parseInt(str,2);
    return decimal;
}


public  void printResults(){
    char uniwavelength = new Character('\u03bb');

    System.out.println("\t\t\t\t\t\t\tA \t\t\t\t\t\t\t\t\t\t\t\t\t\t\tQ  \t\t\t\t\t\t\t Q1 \t\t\t\t\t\t\t\tM");

    for(int i=(RegisterWidthInBits-1); i>=0; i--){
        System.out.print(uniwavelength+""+(i+1)+"["+A[i]+"]");
    }
    System.out.print("\t");

    for(int i=(RegisterWidthInBits-1); i>=0; i--){
        System.out.print(uniwavelength+""+(i+1)+"["+Q[i]+"]");
    }

    System.out.print("\t"+uniwavelength+""+(1)+"["+Q1+"]\t");

    for(int i=(RegisterWidthInBits-1); i>=0; i--){
        System.out.print(uniwavelength+""+(i+1)+"["+M[i]+"]");
```

```java
        }
        System.out.print("\n");
    }


    public int[] getQ(){
        return this.Q;
    }
}
```

# Floating Point Signed Divide

For floating point divide, I set up a project as shown in Fig.B.5

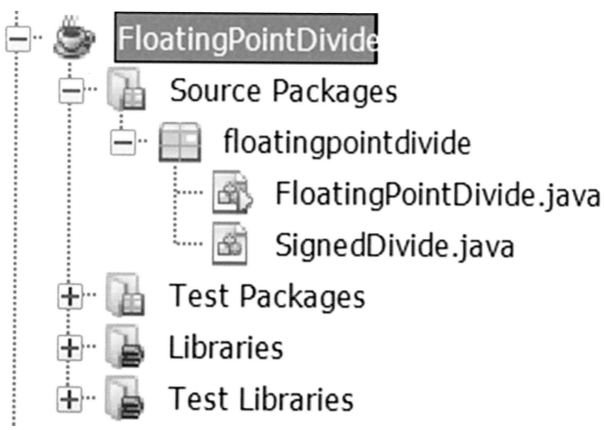

Fig.B.5

I created the Java project with name FloatingPointDivide, and then in this project, I created the classes FloatingPointDivide.java and SignedDivide.java. I typed in the code and highlighted the FloatingPointDivide project. Then I chose on the menu Run and Run Project from the drop-down menu list, and it ran the code.

```
/*
 * Author: Michael Cloran 6/7/19
 * Email: michaelcloran2010@gmail.com
 *
 * Note to user. This code has absolutely no warranty. It works for me on my
 * computer and is here as is and is not deemed as fit for any particular
 * purpose.The author cannot be held liable for damage or loss of data due
 * to its use.The author is not liable for any thing that might go wrong with
 * the codes use.
```

```java
*/ package floatingpointdivide;

/**
 *
 * @author mc201
 * Author: Michael Cloran 19/7/19
 * Email: michaelcloran2010@gmail.com
 * Version 1.0
 * FloatingPointDivide.java
 */
public class FloatingPointDivide {

    private static int RegisterWidthInBits = 32;
    private static int exponentWidthInBits = 8;
    private static int significandWidthInBits = 23;

    private static int Q1 = 0;
    private static int[] A = new int[]{0,0,0,0 ,0,0,0,0, 0,0,0,0, 0,0,0,0, 0,0,0,0, 0,0,0};

    private static int signbitX = 0;
    private static int[] biasedExponentX = new int[] {1,0,0,0, 0,0,0,1};
    private static int[] significandX = new int[]{1,1,1,0 ,0,0,0,0, 0,0,0,0, 0,0,0,0, 0,0,0,0, 0,0,0};

    private static int signbitY = 0;
    private static int[] biasedExponentY = new int[] {1,1,0,0, 0,0,0,0};
    private static int[] significandY = new int[]{1,1,0,0 ,0,0,0,0, 0,0,0,0, 0,0,0,0, 0,0,0,0, 0,0,0};

    private static int signbitZ = 0;
    private static int[] biasedExponentZ = new int[] {0,0,0,0, 0,0,0,0};
    private static int[] significandZ = new int[]{0,0,0,0 ,0,0,0,0, 0,0,0,0, 0,0,0,0, 0,0,0,0, 0,0,0};

    private static int signBit2 = signbitX ^ signbitY;
    /**
```

```
    * @param args the command line arguments
    */
   public static void main(String[] args) {
       if(arrayValuiesEqualsZero(biasedExponentX)==true && arrayValuiesEqualsZero(signifi-
candX)==true && signbitX==0){
           signbitZ = 0;
           for(int i=0; i<exponentWidthInBits;i++ ){
               biasedExponentZ[i] =  0;
           }
           for(int i=0; i<significandWidthInBits;i++ ){
               significandZ[i] =  0;
           }
           printFinalResults();
           return;
       }else{
           if(arrayValuiesEqualsZero(biasedExponentY)==true && arrayValuiesEqualsZero(signifi-
candY)==true && signbitY==0){
               signbitZ = 0;
               for(int i=0; i<exponentWidthInBits;i++ ){
                   biasedExponentZ[i] =  1;
               }
               for(int i=0; i<significandWidthInBits;i++ ){
                   significandZ[i] =  1;
               }
               printFinalResults();
               return;
           }else{
               int exponentX = binaryToInteger(biasedExponentX);
               int exponentY = binaryToInteger(biasedExponentY);
               int exponentZ = 0;

               if(exponentY < 0){
                   exponentZ = (exponentX + exponentY);
```

233

```java
        }else{
            exponentZ = (exponentX - exponentY);
        }


        System.out.println("exponentZ:"+exponentZ+" = exponentX:"+exponentX+ " - expo-
nentY:"+ exponentY);
        exponentZ = (exponentZ + 127);//biased exponent IEEE 754-2008 compatability


        System.out.println("biasedExponentZ:"+ exponentZ);


        String binaryStr = String.format("%8s", Integer.toBinaryString(exponentZ)).replace('
', '0');

        System.out.println("binaryStr:"+binaryStr+ " binaryStrLength:"+ binaryStr.length());


        System.out.print("\nbiasedExponentZ:");
        for(int i=0; i<8;i++){
            biasedExponentZ[i] = Character.getNumericValue(binaryStr.charAt(binaryStr.
length()-1-i));
            System.out.print(biasedExponentZ[i]);
        }
        System.out.println();


        if(exponentZ>255){
            System.out.println("Exponent overflow");
            printFinalResults();
            return;
        }else{
            if(exponentZ<= -255){
                System.out.print("Exponent underflow");
                printFinalResults();
                return;
            }else{


                signbitZ = signbitX;
```

```java
            System.out.print("\nZ:");
            for(int i=0; i<significandWidthInBits; i++){
                significandZ[i] = significandX[i];
                System.out.print(significandZ[i]);
            }
            System.out.println();
            //multiply Significands
            SignedDivide sdObj = new SignedDivide(A,significandZ,significandY);
            int[] temp = sdObj.getQ();
            for(int i=0; i<significandWidthInBits; i++){
                significandZ[i] = temp[i];
            }

            System.out.println("normalizing result");
            normalizeResult();
            System.out.println("Todo round result");

            if(signbitZ != signBit2 ){
                System.out.println("signBit correction");
                signbitZ = signBit2;
                printFinalResults();
            }
            return;
        }
      }
    }
  }
}

public static boolean arrayValuiesEqualsZero(int[] arrayOfValues){
    for(int i=0; i<arrayOfValues.length; i++ ){
        if(arrayOfValues[i] !=0){
```

```java
            return false;
        }
    }
    return true;
}


public static int binaryToInteger(int[] binaryArray){
    String str = "";
    for(int i=(binaryArray.length-1) ; i>= 0; i--){
        str = str+binaryArray[i];
    }


    int decimal = 0;
    for(int i = (binaryArray.length-1); i>=0; i--){
        decimal = decimal + (int)binaryArray[i]*(int)Math.pow(2, i);
    }
    return decimal;
}


public static void normalizeResult(){
    int tempArray[] = new int[32];
        tempArray[31] = signbitZ;
    for(int i =0; i<exponentWidthInBits;i++ ){
        tempArray[i+23] = biasedExponentZ[i];
    }
    for(int i=0; i<significandWidthInBits-1;i++ ){
        tempArray[i] = significandZ[i];
    }


    signbitZ = tempArray[31];


    System.out.println();
    for(int i = (tempArray.length-2); i>=23;i-- ){
```

```java
            biasedExponentZ[i-23] = tempArray[i];
        }
    for(int i=0; i<significandWidthInBits-1;i++ ){
            significandZ[i] = tempArray[i];
        }


    if((signbitZ) == 1 || biasedExponentZ[7] == 1){
        //Todo Round result needed
        System.out.println("Todo round result");
        printFinalResults();
    }else{
        //shifting
        for(int i =tempArray.length-1; i>=0; i--){
            if(i== 0){
                tempArray[i] = 0;


            }else{
                tempArray[i] = tempArray[i-1];
            }
        }


        signbitZ = tempArray[31];
        for(int i =exponentWidthInBits-1; i>23;i-- ){
            biasedExponentZ[i-23] = tempArray[i];
        }
        for(int i=0; i<significandWidthInBits;i++ ){
            significandZ[i] = tempArray[i];
        }


        int tempValue = binaryToInteger(biasedExponentZ);
        tempValue = tempValue -1;


        String binaryStr = String.format("%8s", Integer.toBinaryString(tempValue & 0xFF).
```

```java
replace(' ' , '0');


        for(int i=0; i<8;i++){
            biasedExponentZ[i] = Character.getNumericValue(binaryStr.charAt(binaryStr.
length()-1-i));
        }


        System.out.println("biasedExponentZ:"+binaryToInteger(biasedExponentZ) );


        if((binaryToInteger(biasedExponentZ)<= 0 &&  signbitZ == 1) || (binaryToInteger(bia-
sedExponentZ)<= 0 &&  signbitZ == 0)){
            System.out.println("exponent underflow");
            printFinalResults();
            return;
        }else{
            System.out.println("normalising results");
            printFinalResults();
            normalizeResult();
        }
    }
}


public static void printFinalResults(){
    char uniwavelength = new Character('\u03bb');
    System.out.println("\n\t\tSign \t\t\tEX \t\t\t\t\t\t\t\t\t S");
    System.out.print("InputX \t");
    System.out.print(uniwavelength+""+(32)+"["+signbitX+"]");
    System.out.print("\t");
    for(int i=exponentWidthInBits-1; i>=0;i-- ){
        System.out.print(uniwavelength+""+(32-exponentWidthInBits+i)+"["+biasedExponentX-
[i]+"]");
    }
    System.out.print("\t");
    for(int i=significandWidthInBits-1; i>=0;i-- ){
```

```java
        System.out.print(uniwavelength+""+(32-exponentWidthInBits-significandWidthIn-
Bits+i)+"["+significandX[i]+"]");
        }
        System.out.println("biased form");


        System.out.println();
        System.out.print("InputY \t");
        System.out.print(uniwavelength+""+(32)+"["+signbitY+"]");
        System.out.print("\t");
        for(int i=exponentWidthInBits-1; i>=0;i-- ){
            System.out.print(uniwavelength+""+(32-exponentWidthInBits+i)+"["+biasedExponen-
tY[i]+"]");
        }
        System.out.print("\t");
        for(int i=significandWidthInBits-1; i>=0;i-- ){
            System.out.print(uniwavelength+""+(32-exponentWidthInBits-significandWidthIn-
Bits+i)+"["+significandY[i]+"]");
        }
        System.out.println("biased form");


        System.out.println();
        System.out.print("ResultZ\t");
        System.out.print(uniwavelength+""+(32)+"["+signbitZ+"]");
        System.out.print("\t");
        for(int i=exponentWidthInBits-1; i>=0;i-- ){
            System.out.print(uniwavelength+""+(32-exponentWidthInBits+i)+"["+biasedExpo-
nentZ[i]+"]");
        }
        System.out.print("\t");
        for(int i=significandWidthInBits-1; i>=0;i-- ){
            System.out.print(uniwavelength+""+(32-exponentWidthInBits-significandWidthIn-
Bits+i)+"["+significandZ[i]+"]");
        }
        System.out.println("unbiased form");
    }
```

```java
}

/*
 * Author: Michael Cloran 6/7/19
 * Email: michaelcloran2010@gmail.com
 *
 * Note to user. This code has absolutely no warranty. It works for me on my
 * computer and is here as is and is not deemed as fit for any particular
 * purpose.The author cannot be held liable for damage or loss of data due
 * to its use.The author is not liable for any thing that might go wrong with
 * the codes use.
 */
package floatingpointdivide;

/**
 *
 * @author mc201
 * Author: Michael Cloran 19/7/19
 * Email: michaelcloran2010@gmail.com
 * Version 1.0
 * SignedDivide.java
 */
public class SignedDivide {
                              //LSB           MSB
    private  int[] A;
    private  int[] Q;

    private  int[] M;
    private  int RegisterWidthInBits;
    private  int count;
    private  int signBit;
    /**
```

```java
    * @param args the command line arguments
    */
   public SignedDivide(int[] A,int[] Q,int[] M) {
       this.A = A;
       this.Q = Q;
       this.M = M;

       this.RegisterWidthInBits = this.A.length;
       this.count = this.RegisterWidthInBits;
       this.signBit = Q[this.RegisterWidthInBits-1] ^ M[this.RegisterWidthInBits-1];

       printResults();//printing out the initial state of registers

       while(this.count>0){
           System.out.println("Count:"+this.count);
           System.out.println("==== shifting AQ left ====");

           arithShiftLefttA_Q();

           System.out.println("==== subtracting ====");
           this.A = subtractAmM();
           //Testing for sign bit negative and thus below 0
           if(this.A[this.RegisterWidthInBits-1]==1){
               this.Q[0] = 0;
               System.out.println("==== adding ====");
               this.A = addAmM();
           }else{
               this.Q[0] = 1;
           }

           printResults();
           this.count -= 1;
       }
```

```java
}

public int[] subtractAmM(){
    int num1 = binaryToInteger(this.A);
    int num2 = binaryToInteger(this.M);
    int sum = 0;


    sum = (num1 - num2);
    System.out.println("sum:"+sum+" = num1:"+num1+" - num2:"+num2);
    String binaryStr = String.format("%23s", Integer.toBinaryString(sum)).replace(' ' , '0');
    System.out.println("A:"+binaryStr+" registerWidth:"+this.RegisterWidthInBits+" binaryS-
tringLength:"+binaryStr.length());


    for(int i=0; i<23;i++){
        this.A[i] = Character.getNumericValue(binaryStr.charAt(binaryStr.length()-1-i));
    }


    return this.A;
}

public int[] addAmM(){
    int num1 = binaryToInteger(this.A);
    int num2 = binaryToInteger(this.M);
    int sum = num1 + num2;
    System.out.println("sum:"+sum+" = num1:"+num1+" + num2:"+num2);
    String binaryStr = String.format("%23s", Integer.toBinaryString(sum)).replace(' ', '0');
    System.out.println("A:"+binaryStr+" registerWidth:"+this.RegisterWidthInBits+" binaryS-
tringLength:"+binaryStr.length());


    for(int i=0; i<23;i++){
        this.A[i] = Character.getNumericValue(binaryStr.charAt(binaryStr.length()-1-i));
    }
```

```java
        return this.A;
    }


public void arithShiftLefttA_Q(){
    System.out.println("Shifting A Q left");

    // this is the A registers width (23)+ Q registers width(23) = 46 bits
    int[] AQ = new int[46];//This is a temp register for easy shifting
    System.out.println("length of A:"+this.A.length);

    //copying the values into the temp register
    for(int i=(AQ.length-1); i>=0; --i){
        if(i>=this.RegisterWidthInBits && i<=AQ.length-1 ){
            AQ[i]=this.A[i-(this.RegisterWidthInBits)];
            //System.out.print(AQ[i]);
        }else
        if(i<this.RegisterWidthInBits && i>=0){
            AQ[i] = this.Q[i];
            //System.out.print(AQ[i]);
        }
    }
    //doing the shift A Q left
    for(int i =(AQ.length-1); i>=0; i--){
        if(i== 0){
            AQ[i] = 0;

        }else{
            AQ[i] = AQ[i-1];
        }
    }
    //putting the values back into the registers
    for(int i = 0; i<AQ.length; i++){
        if(i>=this.RegisterWidthInBits && i<AQ.length){
```

```java
                this.A[i-(RegisterWidthInBits)] = AQ[i];
            }else
            if(i<this.RegisterWidthInBits && i>=0 ){
                this.Q[i] = AQ[i];
            }
        }
    }


    public int binaryToInteger(int[] binaryArray){
        String str = "";
        for(int i=(this.RegisterWidthInBits-1) ; i>= 0; i--){
            str = str+binaryArray[i];
        }


        int decimal = 0;
        for(int i = (this.RegisterWidthInBits-1); i>=0; i--){
            decimal = decimal + (int)binaryArray[i]*(int)Math.pow(2, i);
        }
        return decimal;
    }


    public int[] getQ(){
        return this.Q;
    }


    public  void printResults(){
        char uniwavelength = new Character('\u03bb');


        System.out.println("\n\t\t\t\t\t\tA \t\t\t\t\t\t\t\t\t\t\t\tQ    \t\t\t\t\t\t\t\t\t\t\t\tM");


        for(int i=(this.RegisterWidthInBits-1); i>=0; i--){
            System.out.print(uniwavelength+""+(i+1)+"["+this.A[i]+"]");
```

```java
        }
        System.out.print("\t");
        for(int i=(this.RegisterWidthInBits-1); i>=0; i--){
            System.out.print(uniwavelength+""+(i+1)+"["+this.Q[i]+"]");
        }
        System.out.print("\t");
        for(int i=(this.RegisterWidthInBits-1); i>=0; i--){
            System.out.print(uniwavelength+""+(i+1)+"["+this.M[i]+"]");
        }
        System.out.print("\n");
    }
}
```

# Magnitude Comparator

For floating point divide, I set up a project as shown in Fig.B.6

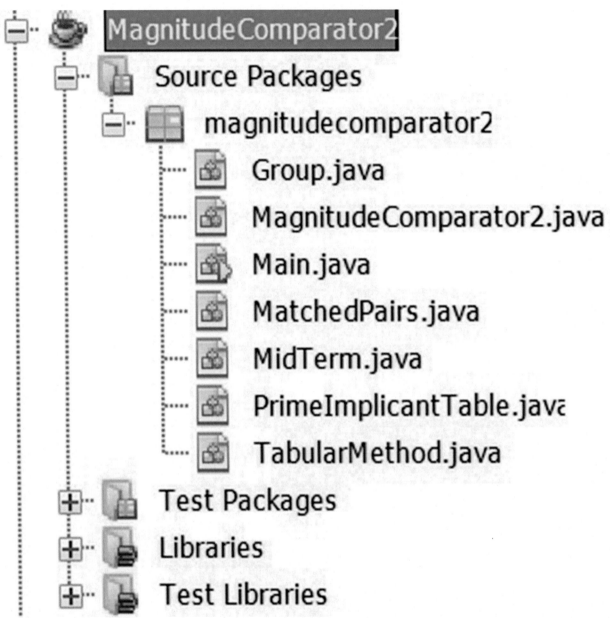

Fig.B.6

I created the Java project with name MagnitudeComparator2 and then, in this project, created the classesGroup.java, MagnitudeComparator2.java, Main.java, MatchedPairs.java, MidTerm. java, PrimeImplicantTable.java, and TabularMethod.java. I typed in the code and highlighted the MagnitudeComparator2 project. Then I chose on the menu Run and Run project from the drop-down menu list, and it ran the code.

```java
/*
 * Author: Michael Cloran 4/9/19
 * Email: michaelcloran2010@gmail.com
 *
 * Note to user. This code has absolutely no warranty. It works for me on my
 * computer and is here as is and is not deemed as fit for any particular
 * purpose.The author cannot be held liable for damage or loss of data due
 * to its use.The author is not liable for any thing that might go wrong with
 * the codes use.
 */
package magnitudecomparator2;

/**
 *
 * @author mc201
 * Author: Michael Cloran 4/9/19
 * Email: michaelcloran2010@gmail.com
 * Version 1.0
 * Main.java
 */
public class Main {

    public static void main(String[] args){
        new MagnitudeComparator2();
    }

}
```

```java
/*
 * Author: Michael Cloran 4/9/19
 * Email: michaelcloran2010@gmail.com
 *
 * Note to user. This code has absolutely no warranty. It works for me on my
 * computer and is here as is and is not deemed as fit for any particular
 * purpose.The author cannot be held liable for damage or loss of data due
 * to its use.The author is not liable for any thing that might go wrong with
 * the codes use.
 */
package magnitudecomparator2;


import java.util.LinkedList;


/**
 *
 * @author mc201
 * Author: Michael Cloran 4/9/19
 * Email: michaelcloran2010@gmail.com
 * Version 1.0
 * Group.java
 */
public class Group {
    private int groupNumber;
    private LinkedList<MidTerm> midTermList = new LinkedList<>();

    public Group(){
        groupNumber = 0;
        midTermList = new LinkedList<>();
    }


    public int getGroupNumber(){
```

```java
        return groupNumber;
    }

    public void setGroupNumber(int number){
        groupNumber = number;
    }

    public void addMidTerm(Integer midTermNumber, boolean checkedValue){
        MidTerm mdObj = new MidTerm(midTermNumber,checkedValue);
        midTermList.add(mdObj);
    }

    public void addMidTerm(Integer midTermNumber,String[] binaryRepArray){
        MidTerm mdObj = new MidTerm(midTermNumber, binaryRepArray);
        midTermList.add(mdObj);
    }

    public void addMidTerm(Integer midTermNumber,String[] binaryRepArray, boolean
checkedValue){
        MidTerm mdObj = new MidTerm(midTermNumber, binaryRepArray, checkedValue);
        midTermList.add(mdObj);
    }

    public void addMidTerm(MidTerm mdTermObj){
        midTermList.add(mdTermObj);
    }

    public LinkedList<MidTerm> getMidTermList(){
        return midTermList;
    }

    public void printGroup(){
        //System.out.println("groupNumber:"+groupNumber);
```

```java
for(MidTerm mt: midTermList){
    System.out.print("\nGroupNumber:"+groupNumber+"\n");
    for(MidTerm.MidTermClass mtClass : mt.getMidTermList()){

        String[] tempBinaryRapArray = mtClass.getBinaryRepArray();
        if(tempBinaryRapArray != null){
            System.out.print(" mt:"+mtClass.getMidTermNumber()+" binaryRep:");
            for(int i =0; i<4; i++){
                if(mtClass.getBinaryRepArray()[i]!=null)System.out.print(mtClass.getBinaryRepArray()[i]);
            }

            if(mtClass.getCheckedBoolean()==true){
                System.out.print("\tTickedBool:\t"+'\u2713');
            }else{
                System.out.print(" \tTickedBool:\t");
            }
            System.out.println();
        }else{
            if(mtClass.getCheckedBoolean()==true){
                System.out.print("\nGroupNumber:"+groupNumber+" mt:"+mtClass.getMidTermNumber()+" \tTickedBool:\t"+'\u2713');
            }else{
                System.out.print("\nGroupNumber:"+groupNumber+" mt:"+mtClass.getMidTermNumber()+" \tTickedBool:\t");
            }
        }
    }
    System.out.println();
}
}
}
```

```java
/*
 * Author: Michael Cloran 4/9/19
 * Email: michaelcloran2010@gmail.com
 *
 * Note to user. This code has absolutely no warranty. It works for me on my
 * computer and is here as is and is not deemed as fit for any particular
 * purpose.The author cannot be held liable for damage or loss of data due
 * to its use.The author is not liable for any thing that might go wrong with
 * the codes use.
 */
package magnitudecomparator2;

import java.util.LinkedList;

/**
 *
 * @author mc201
 * Author: Michael Cloran 4/9/19
 * Email: michaelcloran2010@gmail.com
 * Version 1.0
 * MagnitudeComparator2.java
 */
public class MagnitudeComparator2 {
    private int RegisterWidthInBits = 2;
    public MagnitudeComparator2(){
        RegisterWidthInBits = 2;
        generateTruthTable();
    }

    public void generateTruthTable(){
        int[] A = new int[2];
        int[] B = new int[2];
```

```java
System.out.println("\n\tA2A1\tB2B1 \tA<B\tA=B\tA>B");
int numberCtr=0;
TabularMethod tmObj = new TabularMethod();
TabularMethod tmObj1 = new TabularMethod();
TabularMethod tmObj2 = new TabularMethod();

for(int i=0;i<16;i++){
    String binaryStr = String.format("%4s", Integer.toBinaryString(i & 0xF)).replace(' ',
'0');

    //debug
    //System.out.println("binaryStr:"+binaryStr+" binaryStrLength:"+binaryStr.length());
    String[] tempBinaryRepArray = new String[4];
    for(int x=0;x<4;x++){
        tempBinaryRepArray[x] = ""+(binaryStr.charAt(3-x));
    }
    /*
    System.out.println();
    for(int x=0; x<4;x++){
        System.out.print(tempBinaryRepArray[x]);
    }
    System.out.println();
    */

    int ctr = 0;
    for(int x = 0; x < 2; x++){
        B[1-ctr] = Character.getNumericValue(binaryStr.charAt(3-x));
        ctr += 1;
    }
    for(int x = 0; x < 2; x++){
        A[x] = Character.getNumericValue(binaryStr.charAt(x));
    }
```

252

```java
//debug
/*for(int x=0; x<2;x++){
    System.out.println("A["+A[x]+"] B["+B[x]+"]");
}*/

int[] tempB = new int[2];
int[] tempA = new int[2];
/*for(int x=1;x>=0; x--){
    tempB[x] = B[2-x] ;
    tempA[x] = A[2-x];
}*/
for(int x=1;x>=0; x--){
    tempB[x] = B[1-x] ;
    tempA[x] = A[1-x];
}

//debug
/*for(int x= 0 ; x<4; x++){
    System.out.println("temp:"+tempB[x]+" B:"+binaryToInteger(tempB)+" A:"+binaryTo-
Integer(tempA));
}*/

if(binaryToInteger(tempA) < binaryToInteger(tempB)){
    System.out.print(numberCtr+"\t");
    printArray(A);
    System.out.print("\t");
    printArray(B);
    System.out.print("\t   1\t  0\t   0");
    tmObj.addToGivenMidTerms(i, tempBinaryRepArray);
    numberCtr+=1;
```

```java
        }else
        if(binaryToInteger(tempA) == binaryToInteger(tempB)){
            System.out.print(numberCtr+"\t");
            printArray(A);
            System.out.print("\t");
            printArray(B);
            System.out.print("\t   0\t   1\t   0");
            tmObj1.addToGivenMidTerms(i, tempBinaryRepArray);
            numberCtr++;
        }else{//A>B
            System.out.print(numberCtr+"\t");
            printArray(A);
            System.out.print("\t");
            printArray(B);
            System.out.print("\t   0\t   0\t   1");
            tmObj2.addToGivenMidTerms(i, tempBinaryRepArray);
            numberCtr++;
        }
        System.out.print("\n");
    }


    System.out.println("---- Original Step1 Group ----");
    tmObj2.generateFirstTable();
    System.out.println("---- Original Step1 Group End ----");



    MatchedPairs mpObj = new MatchedPairs(tmObj2.getGroupList());


    System.out.println("---- Original Step2  ----");
    LinkedList<Group> stateOfGroupList = mpObj.generateMatchedPairTable();
    System.out.println("---- End Original Step2 ----");


    System.out.println("---- Original step 1 watch ticks ----");
```

```java
for(Group grp :stateOfGroupList ){
    grp.printGroup();
}
System.out.println("---- End Original Step 1 watch ticks ----");

MatchedPairs mpObj1 = new MatchedPairs(mpObj.getGroupList());

System.out.println("---- Step3 watch ticks ----");
LinkedList<Group> stateOfGroupList2 = mpObj1.generateMatchedPairTable();
System.out.println("\n---- End Step3 ticks ----");

System.out.println("---- Original Step2 Watch ticks ----");
for(Group grp :stateOfGroupList2 ){
    grp.printGroup();
}
System.out.println("---- End Original Step2 Watch ticks ----");

MatchedPairs mpObj2 = new MatchedPairs(stateOfGroupList2);

System.out.println("---- Step4 watch ticks ----");
LinkedList<Group> stateOfGroupList3 = mpObj2.generateMatchedPairTable();
System.out.println("\n---- End Step4 ticks ----");

PrimeImplicantTable piObj = new PrimeImplicantTable();
piObj.addToPrimeImplicantList(tmObj.getGroupList());



piObj.addToPrimeImplicantList(stateOfGroupList);



piObj.addToPrimeImplicantList(stateOfGroupList2);
```

255

```java
        piObj.addToPrimeImplicantList(mpObj2.getGroupList());

        piObj.generateTables();


    }


    public void printArray(int[] binaryArray){
        for(int i =0; i< RegisterWidthInBits; i++){
            System.out.print(binaryArray[i]);
        }
    }


    public  int binaryToInteger(int[] binaryArray){
        String str = "";
        for(int i=(RegisterWidthInBits-1) ; i>= 0; i--){
            str = str+binaryArray[i];
        }
        int decimal = Integer.parseInt(str,2);
        return decimal;
    }
}
```

```
/*
 * Author: Michael Cloran 4/9/19
 * Email: michaelcloran2010@gmail.com
 *
 * Note to user. This code has absolutely no warranty. It works for me on my
 * computer and is here as is and is not deemed as fit for any particular
 * purpose.The author cannot be held liable for damage or loss of data due
 * to its use.The author is not liable for any thing that might go wrong with
 * the codes use.
 */
package magnitudecomparator2;

import java.util.LinkedList;

/**
 *
 * @author mc201
 * Author: Michael Cloran 4/9/19
 * Email: michaelcloran2010@gmail.com
 * Version 1.0
 * MatchedPairs.java
 */
public class MatchedPairs extends TabularMethod{
    private LinkedList<Group> groupList = new LinkedList<>();//??
    private LinkedList<Group> matchedPairGrpList = new LinkedList<>();
    public MatchedPairs(LinkedList<Group> groupLst){
        for(Group grp : groupLst){
            groupList.add(grp);
            //System.out.println("added groupNumber:"+grp.getGroupNumber());
        }
    }
```

```java
public LinkedList<Group> generateMatchedPairTable(){
    for(Group grp : groupList){

        int currentGroupNumber = grp.getGroupNumber();
        int nextGroupNumber = 0;
        Group matchedPairsGrp = new Group();
        nextGroupNumber = currentGroupNumber +1;


        if(nextGroupNumber!=(groupList.size())){

            Group nextGrp = new Group();
                try{
                    nextGrp = groupList.get(nextGroupNumber);
                }catch(IndexOutOfBoundsException iobex){
                    System.out.println("Group n+1 does not exist");
                    //iobex.printStackTrace();
                    nextGrp=null;
                }catch(NullPointerException npex){
                    System.out.println("Group n+1 does not exist");
                    nextGrp = null;
                }


            //matchedPairsGrp = new Group();
            matchedPairsGrp.setGroupNumber(currentGroupNumber);


            for(MidTerm mt : grp.getMidTermList()){
                if(nextGrp != null){
                    for(MidTerm mt2 : nextGrp.getMidTermList()){
                        if(mt.getMidTermList().size()==1){
                            LinkedList<MidTerm.MidTermClass> cObj1 =  mt.getMidTermList();
                            LinkedList<MidTerm.MidTermClass> cObj2 =  mt2.getMidTermList();
                            if(compareAndDetermineIfPositiveAndCanBeExpressedInPowersOfT-
wo(cObj1.getFirst().getMidTermNumber(),cObj2.getFirst().getMidTermNumber()) == true){
```

```
Number(),false);                    MidTerm midtermOBJ = new MidTerm(cObj1.getFirst().getMidTerm-

                                    midtermOBJ.addMidTerm(cObj2.getFirst().getMidTermNumber(),-
false);


                                    midtermOBJ.setBinaryRepArray(mt2.getMidTermList().getFirst().getBi-
naryRepArray());


                                    matchedPairsGrp.addMidTerm(midtermOBJ);


                                    mt.getMidTermList().getFirst().setCheckedBoolean(true);
                                    mt2.getMidTermList().getFirst().setCheckedBoolean(true);


                    }
                }else
                if(mt.getMidTermList().size()==2){


                        int difference1 = mt.getMidTermList().getLast().getMidTermNumber() -
mt.getMidTermList().getFirst().getMidTermNumber();
                        int difference2 = mt2.getMidTermList().getLast().getMidTermNumber() -
mt2.getMidTermList().getFirst().getMidTermNumber();


                        if((difference1 == difference2) && compareAndDetermineIfPositive-
AndCanBeExpressedInPowersOfTwo(mt.getMidTermList().getLast().getMidTermNumber(),mt2.
getMidTermList().getLast().getMidTermNumber())){
                        //if((difference1 == difference2) && ( (checkForTwoToPowerTwo(mt2.
getMidTermList().getLast().getMidTermNumber()-mt.getMidTermList().getLast().getMidTermNum-
ber()) == true) ) ){
                        //System.out.println("checkpow2:"+checkForTwoToPowerTwo(mt2.
getMidTermList().getLast().getMidTermNumber()-mt.getMidTermList().getLast().getMidTermNum-
ber()));
                        MidTerm midtermOBJ = new MidTerm();
                        for(MidTerm.MidTermClass tempMtC : mt.getMidTermList()){
                            tempMtC.setCheckedBoolean(true);
                        }


                        for(MidTerm.MidTermClass mtIntClass : mt.getMidTermList()){
```

```
                    midtermOBJ.addMidTerm(mtIntClass.getMidTermNumber(),false);
                }

                for(MidTerm.MidTermClass tempMtC : mt2.getMidTermList()){
                    tempMtC.setCheckedBoolean(true);
                }

                for(MidTerm.MidTermClass mtIntClass : mt2.getMidTermList()){
                    midtermOBJ.addMidTerm(mtIntClass.getMidTermNumber(),false);
                }

                midtermOBJ.setBinaryRepArray(mt2.getBinaryRepArray());

                System.out.println("testing:"+mt2.printBinaryRepArray());

                matchedPairsGrp.addMidTerm(midtermOBJ);
            }
        }else{//do i need code here for testing for midtermList greater then 2.
            System.out.println("todo testing");
        }
    }
}


    }



    matchedPairGrpList.add(matchedPairsGrp);
    matchedPairsGrp.printGroup();
}
return groupList;//works
//return matchedPairGrpList;
```

```java
        }

    public boolean checkForTwoToPowerTwo(int number){

        if((number % 2) != 0){
            return false;
        }else{
            for(int i=0; i<= number; i++){
                if(Math.pow(2,i) == number){
                    return true;
                }
            }
        }
        return false;
    }


    public boolean compareAndDetermineIfPositiveAndCanBeExpressedInPowersOfTwo(int number1, int number2){
        int difference = number2 - number1;
        if((difference)>1){
            if(checkForTwoToPowerTwo(difference) == true){
                return true;
            }
        }else if(difference == 1){
            return true;
        }
        return false;
    }


    public LinkedList<Group> getGroupList(){
        return matchedPairGrpList;
    }
}
```

```java
/*
 * Author: Michael Cloran 4/9/19
 * Email: michaelcloran2010@gmail.com
 *
 * Note to user. This code has absolutely no warranty. It works for me on my
 * computer and is here as is and is not deemed as fit for any particular
 * purpose.The author cannot be held liable for damage or loss of data due
 * to its use.The author is not liable for any thing that might go wrong with
 * the codes use.
 */
package magnitudecomparator2;

import java.util.LinkedList;

/**
 *
 * @author mc201
 * Author: Michael Cloran 4/9/19
 * Email: michaelcloran2010@gmail.com
 * Version 1.0
 * MidTerm.java
 */
public class MidTerm {

    private LinkedList<MidTermClass> midTermList = new LinkedList<>();
    String primeImplicantTermStr;
    boolean isChecked;
    String[] binaryRepArray;
    int primeImplicantsTableRow = 0;

    public MidTerm(int midTermNumber, String[] binaryRepArray){
        MidTermClass mdClass = new MidTermClass(midTermNumber,binaryRepArray,false);
```

```java
        midTermList.add(mdClass);
    }

    public MidTerm(int midTermNumber, String[] binaryRepArray,boolean checkedValue){
        MidTermClass mdClass = new MidTermClass(midTermNumber,binaryRepArray,checked-
Value);
        midTermList.add(mdClass);
    }

    public MidTerm(int midTermNumber,boolean checkedValue){
        MidTermClass mdClass = new MidTermClass(midTermNumber,checkedValue);
        midTermList.add(mdClass);
    }

    public MidTerm(int midTermNumber){
        MidTermClass mdClass = new MidTermClass(midTermNumber,false);
        midTermList.add(mdClass);
    }

    public MidTerm(){

    }

    public void addMidTerm(int midTermNumber){
        MidTermClass mdClass = new MidTermClass(midTermNumber,false);
        midTermList.add(mdClass);
    }

    public void addMidTerm(int midTermNumber, boolean checkedValue){
        MidTermClass mdClass = new MidTermClass(midTermNumber,checkedValue);
        midTermList.add(mdClass);
    }
```

```java
public LinkedList<MidTermClass> getMidTermList(){
    return midTermList;
}

public int getFirstMidTermNumber(){
    return midTermList.getFirst().getMidTermNumber();
}

public int getSecondMidTermNumber(){
    return midTermList.getLast().getMidTermNumber();
}

public String getPrimeImplicantTermStr(){
    return primeImplicantTermStr;
}

public void setPrimeImplicantTermStr(String primeImplicantTerm){
    primeImplicantTermStr = primeImplicantTerm;
}

public boolean getCheckedBoolean(){
    return isChecked;
}

public void setCheckedBoolean(boolean state){
    isChecked = state;
}

public String printBinaryRepArray(){
    String str = "";
    if(binaryRepArray!=null){
        for(int i =0;i<4;i++){
            str += (binaryRepArray[i]);
```

```
        }
    }
    return str;
}


public String[] getBinaryRepArray(){
    return binaryRepArray;
}


public void setBinaryRepArray(String[] binaryRepArray){
    this.binaryRepArray = binaryRepArray;
}


public int getPrimeImplicantsTableRowNumber(){
    return primeImplicantsTableRow;
}


public void setPrimeImplicantsTableRowNumber(int rowNumber){
    primeImplicantsTableRow = rowNumber;
}


public class MidTermClass extends MidTerm{
    int midTermNumber;
    String[] binaryRepArray = new String[4];
    boolean isChecked;


    public MidTermClass(int number, String[] binaryRepArray,boolean checkedValue){
        midTermNumber= number;
        this.binaryRepArray = binaryRepArray;
        isChecked = checkedValue;
    }
```

```java
public MidTermClass(int number, boolean checkedValue){
    midTermNumber= number;
    isChecked = checkedValue;
}

public MidTermClass(){
    midTermNumber = 0;
}

public int getMidTermNumber(){
    return midTermNumber;
}

public void setMidTermNumber(int number){
    midTermNumber = number;
}

public boolean getCheckedBoolean(){
    return isChecked;
}

public void setCheckedBoolean(boolean state){
    isChecked = state;
}

public String[] getBinaryRepArray(){
    return binaryRepArray;
}

public void setBinaryRepArray(String[] binaryRepArray){
    this.binaryRepArray = binaryRepArray;
}
```

```
    }
}
```

```java
/*
 * Author: Michael Cloran 4/9/19
 * Email: michaelcloran2010@gmail.com
 *
 * Note to user. This code has absolutely no warranty. It works for me on my
 * computer and is here as is and is not deemed as fit for any particular
 * purpose.The author cannot be held liable for damage or loss of data due
 * to its use.The author is not liable for any thing that might go wrong with
 * the codes use.
 */
package magnitudecomparator2;


import java.util.ArrayList;
import java.util.LinkedList;


/**
 *
 * @author mc201
 * Author: Michael Cloran 4/9/19
 * Email: michaelcloran2010@gmail.com
 * Version 1.0
 * PrimeImplicantTable.java
 */
public class PrimeImplicantTable {
    private LinkedList<MidTerm> primeImplicantMidTermList = new LinkedList<>();
                        //LSB                    MSB
    private String[] termsArray = {"B1","B2","A1","A2","E","F","G","H"};
    private String tickMark = ""+'\u2713';
    private ArrayList<String []> columnsArrayList = new ArrayList<>();


    public PrimeImplicantTable(){
```

```java
    }

    public void addToPrimeImplicantList(LinkedList<Group> groupList){
        for(Group grp : groupList){
            for(MidTerm mt : grp.getMidTermList()){
                for(MidTerm.MidTermClass mtC : mt.getMidTermList()){
                    if(mtC.getCheckedBoolean() == false){
                        primeImplicantMidTermList.add(mt);
                        break;
                    }
                }
            }
        }
    }

    public void generateTables(){
        generatePrimeImplicantTermTable();
        generatePrimeImplicantTable();
    }

    public void generatePrimeImplicantTermTable(){
        for(MidTerm mt : primeImplicantMidTermList){
            if(mt.getMidTermList().size() == 1){
                String midtermPrimeImplicantTermStr = "";
                String[] midtermBinaryRepArray = mt.getMidTermList().getFirst().getBinaryRepAr-
ray();

                for(int i=0; i<midtermBinaryRepArray.length; i++ ){
                    if(midtermBinaryRepArray[i] !=null){
                        if(midtermBinaryRepArray[i].equals("1")){
                            midtermPrimeImplicantTermStr += termsArray[i];
                        }else{
                            midtermPrimeImplicantTermStr += termsArray[i]+"'";
```

```java
            }
          }
        }
        mt.setPrimeImplicantTermStr(midtermPrimeImplicantTermStr);
      }else
      if(mt.getMidTermList().size() == 2){
        String midtermPrimeImplicantTermStr = "";
        String[] midtermBinaryRepArray = mt.getBinaryRepArray();


        int difference = mt.getMidTermList().getLast().getMidTermNumber() - mt.getMidTermList().getFirst().getMidTermNumber();
        int underscoreCtr = 0;


        underscoreCtr = getUnderscoreCtr(difference);


        int termsArrayCtr = 0;
        for(int i=0; i<4;i++){
          if(i != underscoreCtr){
            if(midtermBinaryRepArray[i].equals("1")){
              midtermPrimeImplicantTermStr += termsArray[termsArrayCtr];
            }else{
              midtermPrimeImplicantTermStr += termsArray[termsArrayCtr]+"'";
            }
          }
          termsArrayCtr += 1;
        }
        mt.setPrimeImplicantTermStr(midtermPrimeImplicantTermStr);

      }else
      if(mt.getMidTermList().size() == 4){
        String midtermPrimeImplicantTermStr = "";
        String[] midtermBinaryRepArray = mt.getBinaryRepArray();
```

```java
int one = 0;
int two = 0;
int three = 0;
int four = 0;
int ctr = 1;
for(MidTerm.MidTermClass mtC : mt.getMidTermList()){
    if(ctr == 1){
        one = mtC.getMidTermNumber();
    }else
    if(ctr == 2){
        two = mtC.getMidTermNumber();
    }else
    if(ctr == 3){
        three = mtC.getMidTermNumber();
    }else{
        four = mtC.getMidTermNumber();
    }
    ctr += 1;
}


int difference1 = two - one;
int difference2 = four-two;
int underscoreCtr1 = 0;



int underscoreCtr2 = 0;


underscoreCtr1 = getUnderscoreCtr(difference1);
underscoreCtr2 = getUnderscoreCtr(difference2);


System.out.println("underscore1:"+underscoreCtr1+" underscore2:"+underscoreCtr2);

int termsArrayCtr = 0;
```

```java
        for(int i=0; i<4;i++){
            if( (i != underscoreCtr1) && (i != underscoreCtr2)){
                if(midtermBinaryRepArray[i].equals("1")){
                    midtermPrimeImplicantTermStr += termsArray[termsArrayCtr];
                }else{
                    midtermPrimeImplicantTermStr += termsArray[termsArrayCtr]+"'";
                }
            }
            termsArrayCtr += 1;
        }
        mt.setPrimeImplicantTermStr(midtermPrimeImplicantTermStr);


    }else{
        System.out.println("Todo more!!!!");
    }
}
printPrimeImplicantTermTable();


//testing for a duplicate term
if(checkPrineImplicantTermStrAlreadyinTable() == true){
    removePrimeImplicandFromTable();
    System.out.println("++++++++ Corrected Prime Implicant Term Table ++++++++");
    printPrimeImplicantTermTable();
}
}


public void generatePrimeImplicantTable(){

    int midTermCtr = 0;
    for(MidTerm mt: primeImplicantMidTermList){
        midTermCtr+=1;
        mt.setPrimeImplicantsTableRowNumber(midTermCtr);
    }
```

```java
String[] columnArray = new String[(midTermCtr+2)];
columnArray[0] = "-1";
columnsArrayList.add(0, columnArray);//checkMark Column


int midTermCtr2 = 1;
for(MidTerm mt: primeImplicantMidTermList){
    for(MidTerm.MidTermClass mtC: mt.getMidTermList()){
        String[] tempColumnArray = new String[midTermCtr+2];
        for(String row :tempColumnArray){
            row = " ";
        }


        tempColumnArray[0] = ""+mtC.getMidTermNumber();
        if(checkIfAlreadyInList(tempColumnArray[0])==false) columnsArrayList.add(midTermCtr2, tempColumnArray);
    }
    midTermCtr2+=1;
}


for(MidTerm mt : primeImplicantMidTermList){
    int row = mt.getPrimeImplicantsTableRowNumber();
    for(MidTerm.MidTermClass mtC: mt.getMidTermList()){
        for(String[] columnArr : columnsArrayList){
            if(!columnArr[0].equals("-1")){
                String tempStr = ""+mtC.getMidTermNumber();
                if(tempStr.equals(columnArr[0])){
                    columnArr[row] = "X";
                }
            }
        }
    }
}
```

```java
LinkedList<Integer> tempMTList = new LinkedList<>();
int rowNumber = 1;
int rowCtr = 1;


for(String[] columnArr : columnsArrayList){
    int Xctr = 0;
    if(!columnArr[0].equals("-1")){
        for(int row=1;row<columnArr.length;row++){
            if(columnArr[row]!=null){
                if(columnArr[row].equals("X")){
                    Xctr+=1;
                    rowNumber = row;
                }
            }
        }
        if(Xctr == 1){
            columnArr[columnArr.length - 1] = tickMark;
            addTickToRow(columnsArrayList.get(0),rowNumber);
        }else{
            tempMTList.add(Integer.parseInt(columnArr[0]));
        }
    }
}


//todo for checked rows to get the midterms and check the columns for the midtersm
shown in each ticked row.
LinkedList<Integer> untickedColumnsList = getUntickedColumns();

/*for(Integer temp : untickedColumnsList){
    System.out.println("untickedColumnsList:"+temp);
}*/


//tick the column fo rows used
```

```java
String[] tickColumnArr = columnsArrayList.get(0);
for(int row = 1;row < tickColumnArr.length;row++){
    if(tickColumnArr[row]!=null){
        if(tickColumnArr[row].equals(tickMark)){
            for(MidTerm mt: primeImplicantMidTermList){
                if(mt.getPrimeImplicantsTableRowNumber() == row){
                    for(MidTerm.MidTermClass mtC: mt.getMidTermList()){
                        addTickToColumn(mtC.getMidTermNumber());
                        try{
                            if(untickedColumnsList.contains(new Integer(mtC.getMidTermNum-
ber())))untickedColumnsList.remove(new Integer(mtC.getMidTermNumber()));
                        }catch(IndexOutOfBoundsException iobe){
                            System.out.println("untickedColumnsList does not contain "+mtC.
getMidTermNumber());
                        }
                    }
                }
            }
        }
    }
}

/*for(Integer temp : untickedColumnsList){
    System.out.println("1 untickedColumnsList:"+temp);
}*/

//find a row that has midterms columns unticked
LinkedList<Integer> todeleteFromUntickedColumnList = new LinkedList<>();
for(Integer column : untickedColumnsList){
    for(MidTerm mt : primeImplicantMidTermList){
        if(checkIfColumnInUnusedMidTermList(untickedColumnsList,mt)==true){
            for(String[] columnArr : columnsArrayList){
                if(columnArr[0].equals(""+column)){
```

```java
                        addTickToRow(columnsArrayList.get(0),mt.getPrimeImplicantsTable-
RowNumber());

                        addTickToColumn(column);

                        todeleteFromUntickedColumnList.add(column);
                    }
                }
            }
        }
    }


    for(Integer temp : todeleteFromUntickedColumnList){
        if(untickedColumnsList.contains(temp)){
            untickedColumnsList.remove(temp);
        }
    }


    for(Integer temp : untickedColumnsList){
        System.out.println("1 untickedColumnsList:"+temp);
    }


    printPrimeImplicantsTable();


    String expressionStr = "";
    boolean first = true;
    for(MidTerm mt : primeImplicantMidTermList){


        String[] columnTickArr = columnsArrayList.get(0);
        for(int rowNo=1; rowNo<(columnTickArr.length-1);rowNo++){
            if(columnTickArr[rowNo]!=null){
                if(columnTickArr[rowNo].equals(tickMark)){
                    if(mt.getPrimeImplicantsTableRowNumber()== rowNo){
                        if(first==true){

                            expressionStr += mt.getPrimeImplicantTermStr();
```

```java
                    first=false;
                }else{
                    expressionStr += " + "+mt.getPrimeImplicantTermStr();
                }
            }
        }
    }
}
System.out.println("\nResult Expression: "+expressionStr);
}


public void addTickToColumn(int midTermNumber){
    for(String[] columnArr : columnsArrayList){
        if(columnArr[0]!="-1"){
            if(columnArr[0].equals(""+(midTermNumber))){
                System.out.println("Setting tick mark for column:"+columnArr[0]);
                columnArr[columnArr.length-1] = tickMark;
            }
        }
    }
}


public LinkedList<Integer> getUntickedColumns(){
    LinkedList<Integer> untickedColumnsList = new LinkedList<>();
    for(String[] columnArr : columnsArrayList){
        if(!columnArr[0].equals("-1")){
            if(columnArr[columnArr.length-1]!=null){
                if(!columnArr[columnArr.length-1].equals(tickMark)){
                    untickedColumnsList.add(Integer.parseInt(columnArr[0]));
                }
            }else{
                untickedColumnsList.add(Integer.parseInt(columnArr[0]));
```

```java
            }
        }
    }
    return untickedColumnsList;
}


public boolean checkIfColumnInUnusedMidTermList(LinkedList<Integer> untickedColumnsList,MidTerm unusedMidTerm){
    LinkedList<Integer> tempBoolList = new LinkedList<>();
    for(MidTerm.MidTermClass mtC : unusedMidTerm.getMidTermList()){
        if(untickedColumnsList.contains(new Integer(mtC.getMidTermNumber()))){
            tempBoolList.add(1);
        }else{
            tempBoolList.add(0);
        }
    }
    if(unusedMidTerm.getMidTermList().size() == tempBoolList.size()){
        for(Integer tbLNumber : tempBoolList){
            if(tbLNumber != 1 ){
                return false;
            }
        }
    }
    return true;
}


public boolean checkIfMTRowunChecked(MidTerm mt){
    boolean tempBool = false;

        if(checkIfRowUnchecked(mt)==true){
            tempBool = true;
        }
```

279

```java
        return tempBool;
    }


    public boolean checkIfRowUnchecked(MidTerm mt){
        if(columnsArrayList.get(0)[mt.getPrimeImplicantsTableRowNumber()]!=null){
            if(columnsArrayList.get(0)[mt.getPrimeImplicantsTableRowNumber()].equals(tickMark)){
                return false;
            }
        }


        return true;
    }


    public boolean checkIfColumnUnchecked(String[] columnArr){
        if(columnArr[columnArr.length-1]!=null){
            if(columnArr[columnArr.length-1].equals(tickMark)){
                return false;
            }
        }
        return true;
    }


    public boolean checkIfMidTermInTempMTList(LinkedList<Integer> tmpList,MidTerm mt){
        boolean tempBool = false;
        for(MidTerm.MidTermClass mtC: mt.getMidTermList()){
            if(checkIfMTContainsValue(tmpList,mtC.getMidTermNumber())== true){
                tempBool= true;
            }else{
                return false;
            }
        }
        return tempBool;
    }
```

```java
public boolean checkIfMTContainsValue(LinkedList<Integer> tmpList, int number){
    if(tmpList.contains(number) == true){
        return true;
    }
    return false;
}


public boolean checkIfAlreadyInList(String tempColumnArrayZeroStr){
    for(String[] columnArr : columnsArrayList){
        if(columnArr[0] != "-1"){
            if(tempColumnArrayZeroStr.equals(columnArr[0])){
                return true;
            }
        }
    }
    return false;
}


public void printPrimeImplicantsTable(){
    System.out.println("++++++++ Prime Implicants Table ++++++++");
    int rowNumber = 1;
    System.out.print("\t\t");
    for(String[] columnArr : columnsArrayList){
        if(!columnArr[0].equals("-1")) System.out.print("\t"+columnArr[0]);
    }

    for(MidTerm mt : primeImplicantMidTermList){
        int first = 1;
        if(columnsArrayList.get(0)[rowNumber] != null){
            System.out.print("\n"+columnsArrayList.get(0)[rowNumber]+" "+mt.getPrimeImpli-
cantTermStr()+" ");
        }else{
```

```java
        System.out.print("\n"+""+mt.getPrimeImplicantTermStr()+" ");
    }
    for(MidTerm.MidTermClass mtC: mt.getMidTermList()){
        System.out.print(mtC.getMidTermNumber()+",");
    }
    System.out.print("|");
    for(String[] columnArr : columnsArrayList){

        if(columnArr[rowNumber]!=null && columnArr[rowNumber]!=tickMark){
            if(first == 1&& mt.getMidTermList().size()!=4){
                System.out.print("\t"+columnArr[0]+columnArr[rowNumber]);

            }else if(mt.getMidTermList().size()==4 && first ==1){
                first = 2;
                System.out.print(""+columnArr[0]+columnArr[rowNumber]);
            }else{
                System.out.print("\t"+columnArr[0]+columnArr[rowNumber]);
            }
        }else{


                System.out.print("\t ");


        }
    }
    System.out.println("\n");
    rowNumber += 1;
}
System.out.print("\n\t");
for(String[] columnArr : columnsArrayList){

    if(columnArr[columnArr.length-1] != null){
        System.out.print("\t"+columnArr[0]+columnArr[columnArr.length-1]);
    }else{
```

```java
            System.out.print("\t");
        }
    }
    System.out.println();


}


public void printPrimeImplicantTermTable(){
    System.out.println("---- Prime Implicants ----\n Decimal\tBinary\tTerm");
    for(MidTerm mt : primeImplicantMidTermList){
        for(MidTerm.MidTermClass mtC : mt.getMidTermList()){
            System.out.print(mtC.getMidTermNumber()+",");
        }
        if(mt.getMidTermList().size()==2){
            System.out.print("("+(mt.getMidTermList().getLast().getMidTermNumber()-mt.getMid-
TermList().getFirst().getMidTermNumber())+")");
        }else
        if(mt.getMidTermList().size()==4){
            int one = 0;
            int two = 0;
            int three = 0;
            int four = 0;
            int ctr = 1;
            for(MidTerm.MidTermClass mtC : mt.getMidTermList()){
                if(ctr == 1){
                    one = mtC.getMidTermNumber();
                }else
                if(ctr == 2){
                    two = mtC.getMidTermNumber();
                }else
                if(ctr == 3){
                    three = mtC.getMidTermNumber();
                }else{
```

283

```java
            four = mtC.getMidTermNumber();
        }
        ctr += 1;
    }
    System.out.print("("+(two-one)+","+(four-two)+")");
}


    System.out.print("\t"+mt.printBinaryRepArray()+"\t"+ mt.getPrimeImplicantTermStr());
    System.out.println();


    }


}




public void addTickToRow(String[] columnArr,int rowNumber){
    columnsArrayList.get(0)[rowNumber] = tickMark;
}

public void removePrimeImplicandFromTable(){
    LinkedList<MidTerm> tempMidTermList = new LinkedList<>();
    LinkedList<MidTerm> tempToRemoveList = new LinkedList<>();

    for(MidTerm mt : primeImplicantMidTermList){
        String tempPIStr = (mt.getPrimeImplicantTermStr());
        if(tempMidTermList.contains(getMidTermAlreadyInList(tempPIStr))==false){
            tempMidTermList.add(getMidTermAlreadyInList(tempPIStr));
        }else{
            //primeImplicantMidTermList.remove(mt);
            tempToRemoveList.add(mt);
            System.out.println("Duplicate in list:"+getMidTermAlreadyInList(tempPIStr).getPrimeImplicantTermStr());
        }
```

```
        }

        for(MidTerm tempMT: tempToRemoveList){
            primeImplicantMidTermList.remove(tempMT);
        }
    }


    public MidTerm getMidTermAlreadyInList(String tempPIStr){
        for(MidTerm mtnext : primeImplicantMidTermList){
                if(tempPIStr.equals(mtnext.primeImplicantTermStr)){
                    return mtnext;
                }
            }
        return null;
    }


    public boolean checkPrineImplicantTermStrAlreadyinTable(){
        for(MidTerm mt : primeImplicantMidTermList){
            String tempPIStr = (mt.getPrimeImplicantTermStr());
            for(MidTerm mtnext : primeImplicantMidTermList){
                if(tempPIStr.equals(mtnext.primeImplicantTermStr)){
                    return true;
                }
            }
        }

        return false;
    }


    public int getUnderscoreCtr(int difference){
        int underscoreCtr = 0;

        String differenceBinaryStr = String.format("%4s", Integer.toBinaryString(difference &
```

```java
0xFF)).replace(' ', '0');
    String[] differenceArray = new String[8];

    for(int x=0;x<4;x++){
        differenceArray[x] = ""+(differenceBinaryStr.charAt(3-x));
    }

    for(int i=0;i<4;i++){
        if(differenceArray[i].equals("1")){
            underscoreCtr = i;
            break;
        }
    }
    return underscoreCtr;
    }


}
```

```
/*
 * Author: Michael Cloran 4/9/19
 * Email: michaelcloran2010@gmail.com
 *
 * Note to user. This code has absolutely no warranty. It works for me on my
 * computer and is here as is and is not deemed as fit for any particular
 * purpose.The author cannot be held liable for damage or loss of data due
 * to its use.The author is not liable for any thing that might go wrong with
 * the codes use.
 */
package magnitudecomparator2;

import java.util.LinkedList;

/**
 *
 * @author mc201
 * Author: Michael Cloran 4/9/19
 * Email: michaelcloran2010@gmail.com
 * Version 1.0
 * TabularMethod.java
 */
public class TabularMethod {
    private LinkedList<Group> groupList = new LinkedList<>();
    private LinkedList<MidTerm> givenMidTermsList = new LinkedList<>();
    private int numberOfBits = 8;

    public TabularMethod(){

    }

    public void generateFirstTable(){
```

```java
    for(int i = 0; i<=4; i++){
        Group group = new Group();
        group.setGroupNumber(i);
        getMidtermFromGivenMidtermsWithGroup(group,i);
    }


}


public void  getMidtermFromGivenMidtermsWithGroup(Group group, int numberOfOnes){
    int tempNumberOfOnes = 0;

    for(MidTerm mt : givenMidTermsList){
        for(MidTerm.MidTermClass mtClass : mt.getMidTermList()){
            String[] tempBinaryRepArray = mtClass.getBinaryRepArray();
            for(int i=0 ; i<4; i++){
                if(tempBinaryRepArray[i].equals("1")){
                    tempNumberOfOnes++;
                }
            }

            if(tempNumberOfOnes == numberOfOnes){

                //for(MidTerm tempMt : group.getMidTermList()){
                    //System.out.println("here 4");
                    group.addMidTerm(mt);
                //}

            }
            tempNumberOfOnes=0;
        }
    }
    group.printGroup();
    //if(group!=null){
```

```java
        groupList.add(group);
    //}else{
    //    System.out.println("group == null");
    //}


    }


    public void primtImplicantTermDeterminationTable(LinkedList<Group> grp){
        //todo prime implicant term determination table
    }


    public void generatePrimeImplicantTable(LinkedList<Group> grp){


    }


    public LinkedList<MidTerm> getGivenMidTerms(){
      return givenMidTermsList;
    }


    public LinkedList<Group> getGroupList(){
       return groupList;
    }


    public void generatePrimeImplicantTable(){


    }


    public void addToGivenMidTerms(int number, String[] binaryRepArray){
        MidTerm mdObj = new MidTerm(number, binaryRepArray);
        givenMidTermsList.add(mdObj);
    }
}
```

# References

## 1.1.Introduction

[1] *A Treatise on Electricity and Magnetism* by James Clerk Maxwell

[2] *Optical Fiber Communications, 3rd Edition*, by Gerd Keiser.

[3] *Introduction to Electrodynamics, 3rd Edition*, by David J. Griffiths.

[4] *Maxwell's Equations and the Principles of Electromagnetism* by Richard Fitzpatrick

[5] *Understanding Optics with Python* by Vasudevan Lakshminarayanan, Hassen Ghalila, Ahmed Ammar, and L. Srinivasa Varadharajan

## 1.2.Optical Switches

[1] *Optical Waves in Layered Media* by Pochi Yeh

[2] *Photonic Crystals: Mathematical Analysis and Numerical Approximation* by Willy Dörfler, Armin Lechleiter, Michael Plum, Guido Schneider, and Christian Wieners.

[3] *Photonic Crystals: Moulding the Flow of Light, 2nd Edition*, by John D. Joannopoulos, Steven G. Johnson, Joshua N. Winn, and Robert D. Meade.

[4] *Fundamentals of Photonic Crystal Guiding* by Maksim Skorobogatiy and Jianke Yang.

[5] *Light Propagation in Gain Media Optical Amplifiers* by Malin Premaratne and Govind P. Agrawal.

[6] *Nonlinear Optics and Photonics* by Guang S. He.

[7] *Fourier Modal Method and Its Applications in Computational Nanophotonics* by Hwi Kim, Junghyun Park, and Byoungho Lee.

[8] *Nonlinear Optics, 3rd Edition*, by Robert W. Boyd.

[9] *Nonlinear Optics* by E. G. Sauter

[10] *Photonic Crystals Towards Nanoscale Photonic Devices, 2nd Edition*, by Henri Benisty, Vincent Berger, Jean-Michael Gérard, Daniel Maystre, and Alexis Tchelnokov, with contribution by Dominique Pagnoux.

[11] *Finite Element Modeling Methods for Photonics* by B. M. Azizur Rahman and Arti Agrawal.

[12] *Semiconductor Optical Amplifiers* by Michael J. Connelly

[13] *Optical Solitons: From Fibers to Photonic Crystals* by Yuri S. Kivshar, Govind P. Agrawal.

[14] *Photonic Crystals: The Road from Theory to Practice* by Steven G. Johnson, John D. Joannopoulos.

[15] *Photonic Crystals: Theory, Applications, and Fabrication* by Dennis W. Prather, Shouyuan Shi, Ahmed Sharkawy, Janusz Murakowski, Garrett J. Schneider.

[16] *Introduction to Optical Waveguide Analysis: Solving Maxwell's Equations and the Schrödinger Equation* by Kenji Kawano, Tsutomu Kitoh.

[17] *Optical Bistability: Controlling Light with Light* by Hyatt M. Gibbs.

[18] *Nonlinear Optical Systems: Principles, Phenomena, and Advanced Signal Processing* by Le Nguyen Binh, Dang Van Liet.

[19] *Ultra-Fast Fiber Lasers: Principles and Applications with MATLAB Models* by Le Nguyen Binh and Nam Quoc Ngo.

[20] *Physics of Photonic Devices, 2nd Edition*, by Shun Lien Chuang.

[21] *Nonlinear Optical Systems* by Luigi Lugiato, Franco Prati, and Massimo Brambilla.

[22] *Computational Photonics* by Salah Obayya.

[23] *Optical Fiber Communications Systems: Theory and Practice with MATLAB and Simulink Models* by Le Nguyen Binh.

[24] *Principles of Nano-Optics* by Lukas Novotny and Bert Hecht.

[25] *Introduction to Fourier Optics, 3rd Edition*, by Joseph W. Goodman.

[26] *Optics, 4th International Edition*, by Eugene Hecht.

[27] *Understanding Fiber Optics, 5th Edition*, by Jeff Hecht.

[28] *Quantum Optics* by Girish S. Agarwal.

[29] *Practical Applications of Microresonators in Optics and Photonics* by Andrey B. Matsko.

[30] *Slow Light: Science and Applications* by Jacob B. Khurgin and Rodney S. Tucker.

[31] *Optical Switches: Materials and Design* by Baojun Li and Soo Jin Chua.

# Processor Functions / Optical Core Theory

[1] *Digital Design and Computer Organization* by Hassan A. Farhat.

[2] *Logic and Computer Design Fundamentals, 5th Edition*, by Morris Mano, Charles R. Kime, and Tom Martin.

[3] *Computer System Architecture, 3rd Edition*, by M. Morris Mano.

[4] *Digital Design, 2nd Edition*, by M. Morris Mano.

[5] Learn about Electronics. 'Digital Electronics' (accessed 18/6/19), http://www.learnabout-electronics.org/dig53.php.

[6] *Digital Computer Electronics* by Malvino Brown.

[7] *Wikipedia, the Free Encyclopedia*, s.v. 'Adder (electronics)' (accessed July 1, 2019), https://en.wikipedia.org/wiki/Adder_(electronics).

[8] *Computer Organization & Architecture Designing for Performance, 7th Edition*, by William Stallings.

[9] *Design of Arithmetic Units for Digital Computers* by John B. Gosling

[10] *Digital Arithmetic* by Miloš D. Ercegovac and Tomás Lang.

[11] *IA-64 and Elementary Functions: Speed and Precision* by Peter Markstein.

[12] *Handbook of Floating-Point Arithmetic, 2nd Edition*, by Jean-Michel Muller, Nicolas Brunie, Florent de Dinechin, Claude-Pierre Jeannerod, Mioara Joldes, Vincent Lefévre, Guillaume Melquiond, Nathalie Revol, and Serge Torres.

[13] *Computer Arithmetic Algorithms, 2nd Edition*, by Israel Koren.

[14] *Computer Arithmetic: Algorithms and Hardware Designs, 2nd Edition*, by Behrooz Parhami.

[15] *What Every Computer Scientist Should Know About Floating-Point Arithmetic* (accessed 18/7/29), https://docs.oracle.com/cd/E19957-01/806-3568/ncg_goldberg.html.

[16] IEEE 754™-2008—IEEE Standard for Floating-point Arithmetic by IEEE Computer Society. Sponsored by the Microprocessor Standards Committee.

[17] *Modified Quine-McCluskey Method: Fun with Logic Minimization* by Vitthal B. Jadhav.

[18] *Microprocessor Architecture: From Simple Pipelines to Chip Multiprocessors* by Jean-Loup Baer.

[19] *Computer Organisation and Design, RISC-V Edition: The Hardware/Software Interface* by David A. Patterson and John L. Hennessy.

[20] *Neural and Massively Parallel Computers: The Sixth Generation* by Branko Souček and Marina Souček.

[21] *Scalable Shared-Memory Multiprocessing* by Daniel E. Lenoski and Wolf-Dietrich Weber.

[22] *Advanced Multicore Systems-On-Chip: Architecture, On-Chip Network, Design* by Abderazek Ben Abdallah.

[23] *But How Do It Know? The Basic Principles of Computers for Everyone* by J. Clark Scott.

[24] *Vector and Parallel Processors in Computational Science: Proceedings of the Second International Conference on Vector and Parallel Processors in Computational Science, Oxford, 28–31 August 1984* by I. S. Duff and J. K. Reid.

[25] *Advanced Computer Architecture and Parallel Processing* by Hesham El-Rewini and Mostafa Abd-El-Barr.

[26] *Multiprocessor System Architectures: A Technical Survey of Multiprocessor/Multithreaded Systems Using SPARC®, Multi-level Bus Architectures and Solaris® (SunOS™)* by Ben Catanzaro.

[27] *Computer Architecture: A Quantitative Approach, 4th Edition*, by John L. Hennessy and David A. Patterson.

[28] *RISC Architectures* by J. C. Heudin and C. Panetto.

[29] *Wikipedia, the Free Encyclopedia*, s.v. 'Operand forwarding' (accessed August 31, 2019), https://en.wikipedia.org/wiki/Operand_forwarding.

[30] *Stream Processor Architecture* by Scott Rixner, foreword by Bill Dally.

[31] *A Pipelined Multi-core MIPS Machine: Hardware Implementation and Correctness Proof* by Mikhail Kovalev, Silvia M. Müller, and Wolfgang J. Paul.

[32] *Processor Architecture: From Dataflow to Superscalar and Beyond* by Jurij Šilc and Borut Robič and Theo Ungerer.

[33] *Computer Organization and Design, ARM® Edition: The Hardware/Software Interface* by David A. Patterson and John L. Hennessy.

[34] *Computer Organization and Design, 4th Edition: The Hardware/Software Interface* by David A. Patterson and John L. Hennessy.

[35] *Intel® Xeon Phi™ Coprocessor Architecture and Tools: The Guide for Application Developers* by Rezaur Rahman.

[36] *Multiprocessor Methods for Computer Graphics Rendering* by Scott Whitman.

[37] *Fundamentals of Parallel Multicore Architecture* by Yan Solihin.

[38] *Multicore Processors and Systems* by Stephen W. Keckler, Kunle Olukotun, and H. Peter Hofstee.

[39] *Superscalar Microprocessor Design* by Mike Johnson.

[40] *Memory Systems: Cache, DRAM, Disk* by Bruce Jacob, Spencer W. Ng, and David T. Wang.

[41] *Storage Networks Explained: Basics and Application of Fibre Channel SAN, NAS, iSCSI, InfiniBand, and FCoE* by Ulf Troppens, Rainer Erkens, Wolfgang Müller-Friedt, Rainer Wolafka, and Nils Haustein.

## 2.1.1.Instruction Set Listing for RISC-V

[1] *The RISC-V Instruction Set Manual, Volume I: User-Level ISA, Document Version 20190305-Base-Ratification*, editors Andrew Waterman and Krste Asanovic, RISC-V Foundation, March 2019.

[2] *Computer Organisation and Design, RISC-V Edition: The Hardware/Software Interface* by David A. Patterson and John L. Hennessy.

[3] *The RISC-V Reader: An Open Architecture Atlas* by David Patterson and Andrew Waterman.

## 3.2.Basic Volumetrics

[1] *About 3D Volumetric Displays* by Barry G. Blundell.

[2] *Enhanced Visualization: Making Space for 3-D Images* by Barry G. Blundell.

[3] *Volumetric Three Dimensional Display Systems* by Barry Blundell and Adam Schwarz.

## 4.1.2.3.Terrain

[1] *Metamaterials: Critique and Alternatives* by Ben A. Munk.

[2] *Metamaterials with Negative Parameters: Theory, Design, and Microwave Applications* by Ricardo Marqués, Ferran Martín, and Mario Sorolla.

[3] *Waves in Metamaterials* by Laszlo Solymar and Ekaterina Shamonina.

## 4.1.2.6.Different Materials

[1] *Wikipedia, the Free Encyclopedia*, s.v. 'Bark (botany)' (accessed October 27, 2018), https://en.wikipedia.org/wiki/Bark_(botany).

## 4.1.2.8.Different Colours

[1] *Wikipedia, the Free Encyclopedia*, s.v. 'Visible spectrum' (accessed October 27, 2018), https://en.wikipedia.org/wiki/visible_spectrum.

## 4.3.1. Basic Laws of Physics

[1] ThoughtCo, 'Introduction to the Major Laws of Physics' (accessed November 18, 2018), https://www.thoughtco.com/major-laws-of-physics-2699071.

[2] *The Principia: Mathematical Principles of Natural Philosophy* by Isaac Newton, translated by I. Bernard Cohen and Anne Whitman, assisted by Julia Budenz.

[3] *Applied Mechanics, 2nd Edition*, by J Hannah and MJ Hillier.

[4] *Classical Mechanics, 3rd Edition*, by Herbert Goldstein, Charles P. Poole, and John Safko.

[5] *Advanced level physics, 6th Edition*, by Michael Nelkon and Philip Parker.

[6] *Wikipedia, the Free Encyclopedia*, s.v. 'Newton's laws of motion' (accessed November 18, 2018), https://en.wikipedia.org/wiki/Newton%27s laws of motion.

[7] *The Feynman Lectures on Physics, Volumes I, II, III* by Richard P. Feynman, Robert B. Leighton, and Matthew Sands.

[8] *A Student's Guide to Einstein's Major Papers* by Robert E. Kennedy.

## 4.3.2. Basic Airflow/Wind

[1] *Fluid Engine Development* by Doyub Kim.

[2] *A First Course in Computational Fluid Dynamics* by H. Aref and S. Balachandar.

[3] *Aerodynamics for Engineers International, 6th Edition*, by John J. Bertin and Russell M. Cummings.

[4] *Fluid Mechanics: First SI Metric Edition* by Streeter Wylie.

[5] *Fluid Mechanics Demystified* by Merle C. Potter.

[6] Free Science Lessons, 'Forces Acting on a Skydiver' (accessed December 8, 2018), https://www.freesciencelessons.co.uk/gcse-physics-paper-2/forces/.

### 4.3.5. Character Modelling

[1] *Inspired 3D Modeling and Texture Mapping* by Tom Capizzi.

### 4.3.6. Rigging a Character in Volumetric Space

[1] *Animation Methods—Rigging Made Easy! Rig Your First 3D Character in Maya* by David Rodriguez.

[2] *Essential Skills in Character Rigging* by Nicholas B. Zeman.

[3] *Inspired 3D Advanced Rigging and Deformations* by Brad Clark, John Hood, and Joe Harkins.

[4] *Inspired 3D Character Setup* by Michael Ford, Alan Lehman.

### 4.3.7. Animation.

[1] *Animation Methods: Becoming a 3D Character Animator* by David Rodriguez.

[2] *Inspired 3D Character Animation* by Kyle Clark.

Printed in the United States
By Bookmasters